A LONGING FOR LAND

To Pia and Shallum

A Longing for Land

Tradition and change in a Swazi agricultural community

PAUL N. BOWEN

Avebury

Aldershot · Brookfield USA · Hong Kong · Singapore · Sydney

Published by
Avebury
Ashgate Publishing Limited
Gower House
Croft Road
Aldershot
Hants GU11 3HR
England

Ashgate Publishing Company
Old Post Road
Brookfield
Vermont 05036
USA

British Library Cataloguing in Publication Data

Bowen, Paul N.
Longing for Land: Tradition and Change in a Swazi Agricultural Community
I. Title
307.72096887

ISBN 1 85628 398 4

Printed and Bound in Great Britain by
Athenaeum Press Ltd, Newcastle upon Tyne.

Contents

List of figures	vi
Preface	vii
1 Swaziland: History and review	1
2 The Ntimane Irrigation Scheme	31
3 Land tenure	57
4 Labour and finance	88
5 Kinship, lineage and locality	111
6 Power and authority	139
7 Social differentiation	169
8 Real Farmers, society and the state	188
Bibliography	217
Index	236

List of figures

Figure 1.1	Southern Africa	3
Figure 1.2	Swaziland	7
Figure 1.3	Land tenure in Swaziland	8
Figure 1.4	Swazi government structure	11
Figure 2.1	Ntimane Irrigation Scheme	38
Figure 2.2	Ntimane Irrigation Scheme: Major stages of development 1966-1986	39
Figure 5.1	A "traditional" Swazi homestead	121
Figure 5.2	An intermediate homestead	124
Figure 6.1	The Chief's administration in Ngwempisi	141

Preface

Over the last 30 years many countries have instituted development programmes aimed at the rural economy. In Swaziland the Rural Development Areas Programme (RDAP) was intended as the means by which agricultural production could be increased so as to redress problems of land scarcity and overpopulation in rural areas. These goals were never achieved and government enthusiasm for the scheme gradually waned.

This book documents and explains the unintended results of a major development scheme. Researched and written from the perspective of social anthropology, this case study describes processes of sociocultural change which go beyond often unquestioned notions of class and class differentiation.

The protagonists are people who are seeking proprietorial control over the land which they currently use by right of usufruct. Their endeavours take the form of intense political activity in local affairs, in which they support certain customary institutions, while striving to achieve their own long term goal of accumulating agricultural land. The resultant social changes are accepted as valid by the majority of the community. The process of change in the area is also part of general change occurring in Swaziland, and there are many similarities apparent between the manner in which some people support tradition, and the way in which the Swazi state has evolved over the last twenty years.

I conducted the research on which this book is based for my doctoral thesis in 1985 and 1986. The research focuses on a cooperative vegetable growing organisation I have pseudonymously called the Ntimani Irrigation Scheme, located in a chiefdom in the mid-west of Swaziland.

In preparing this monograph, I would like to re-acknowledge the contribution made to the original fieldwork by many people both in Australia and in Swaziland. I am particularly indebted, as always, to my wife Dr Pia Broderick for the many reasons only she can understand.

1 Swaziland: History and review

Introduction

The configuration of local communities in Swaziland and their relationship to the Swazi state have always rested on the use of land. Land was the basis of royal power in the pre-colonial period, and social organization in the chiefdoms was determined by the manner in which land was acquired and used. This is a book about social change and the complex interaction between the use of land and the nature of social organization in Swaziland. This introductory chapter is in three parts. First is a brief historical overview of the process by which land, used under traditional Swazi systems of production, became scarce. Second is a description of contemporary social organization and an introduction to the farmers who are the focus of this book. Within this, a review of the perceptions of Swazi rural society as portrayed in recent literature is given. Finally, is a description of the Rural Development Areas Programme and the state of contemporary rural development policies in Swaziland.

A brief history of the Swazi

The early history of the Swazi[1] has been well documented in various authoritative histories (Bonner, 1983; Booth, 1983; Kuper, 1941, 1947, 1978; Matsebula, 1980, 1983). The Swazi are descendants of the Nguni speaking peoples who came southward from east central Africa, crossed the Limpopo River and settled in Tsongaland (Mozambique) in the late fifteenth century. Their leader was a man named Dlamini, who was the founder of the royal

Swazi clan (Booth, 1983, p.7; Kuper, 1947, p.11). In the late eighteenth century, the Swazi first moved into the southern part of modern day Swaziland (see Figure 1.1). Subsequently, during the *mfecane*,[2] under Sobhuza I, the Swazi moved north, simultaneously assimilating other Nguni and Sotho clans as they went. The Swazi established themselves in country with clear evidence of both prior occupation, dating back to prehistoric times,[3] and current use by the Khoi-san (Bushmen) (Jones, 1981, p.349; Masson, 1982, p.7; Matsebula, 1980, p.2).

By 1820, the Swazi king was established at Lobamba, the site of the present royal capital, and the Swazi empire was spreading into the Transvaal. By 1865, Swazi military and administrative organization, based heavily on the Ndwandwe Age Regiment system and legitimated through the annual Ndwandwe *incwala* (first fruits) ceremony, was firmly established. By this time, the Swazi had become of strategic interest to the British and to the Boer Republics, and the Swazi eventually gained a reputation for securing their territory by indirectly supporting one or other of the greater powers, through direct and formal alliances.

In the late 1880s, Swazi sovereignty was traded by the European powers as promises were made and broken. After the Second Boer War, an unsympathetic British administration, under Lord Milner (Pakenham, 1979, p.119), assumed control of Swaziland. By 1907, extensive land grants made during the 'Concessions Period' (1880-1900) were ratified by the British colonial government through their new Resident Commissioner, who was stationed at the town of Mbabane with a full administrative staff. The traditional structure of government was maintained under these arrangements, although ultimate authority rested with the Commissioner.

In 1921, Sobhuza II became *Ngwenyama* (King).[4] While British power was consolidated at this time, Sobhuza led Swazi demands for independence. This period of Swazi history was marked by very high taxation and growing settler power. With the advent of World War II, Britain's policy of neglect, in terms of capital investments, changed and various mineral processing and agricultural industries were established, resulting in a great demand for labour. Indirect rule remained in force in Swaziland until the elections of 1963 and, in 1968, Sobhuza II became head of state under a Westminster system of government. The next five years were turbulent, and in 1973, Sobhuza II seized power as an absolute monarch. He consolidated this position over the next decade, until his death on 21st August 1982.

Figure 1.1 Southern Africa

He left a fifteen year old heir, Prince Makhositive, in the care of Queen Regent Dzeliwe, the fifteen member *Liqoqo* (Inner Council) and Prince Sozisa, 'the Authorised Person', who was a member of the royal family appointed by Sobhuza, prior to his death, to assist the Queen Regent. After Sobhuza's death there was a period of unrest, centring on the machinations of Makhositive's uncle, Prince Mfanasibili. This came to an end with the coronation of Makhositive, as King Mswati III, in 1986.

Traditional Swazi society

Swaziland has been a stratified society, involving an aristocracy and commoners, for over one hundred years. The aristocracy consisted of members of the ruling Dlamini clan, and also included other hereditary chiefs. This hierarchy was sustained by its particular relationship with the process of production. The relative position of individuals within the hierarchy was largely dependent upon clan and lineage affiliations.

Throughout the nineteenth and into the twentieth century, systems of production and social organization evolved largely in accordance with the demands of the social and physical environment. At the core of the Swazi economic system was the right of a married man to *khonta* (to give allegiance) to a chief, in return for which he gained the right to build his own *umuti* (homestead) within the *sive* (chiefdom). These rights could not be achieved by unmarried men, and were achieved at a particular stage in the developmental cycle of the homestead. After marriage, a man may have stayed with his wife (or wives) and children, as part of the homestead of which his father or eldest brother was headman. Therefore, he did not give allegiance until he left this homestead and set up his own homestead on land that was not within the existing homestead's land. At this point he became a headman in his own right. The homestead, a configuration of huts, was the dwelling place of a man and his wives and children, and was based on the *indhlu* (house) as its most basic unit. This form of organization is known as the house/property complex, and represents a form of organization common to Southern Africa.

According to Gluckman (1950), the house property complex was a system whereby:

> ... a woman transmits to her sons specific rights in their patrimony, and no rights in her own family property ... a woman passes rights in the patrimony to her sons against their father's sons by other wives (1950, p.197).

The 'house' comprised a woman and her children, under the judicial authority of a headman, and the administrative authority of his mother (Kuper 1950, p.92). 'After a period of service to her mother-in-law each wife is ... allotted her own fields, and, if possible, cattle for her use' (Kuper, 1963, p.20). In time, this house formed the nucleus of a new homestead (Schapera, 1966, p.59; Sheddick, 1953, p.63), at which point a woman would achieve her ultimate status as *gogo* (grandmother) in the homestead of her eldest son, where '... she takes charge of his domestic affairs, and lives in the great hut [of her son's homestead]' (Kuper, 1950, p.95). This status was thus enhanced by the woman's access to land, which she retained while she used it. This right to land reinforced the focal role women occupied, particularly in traditional society, in sustaining the rural base of each family group.

The rural economy was subsistence based, dominated by cultivation and sedentary pastoralism, with minimal hunting. Cultivation provided mostly maize, but also some vegetables (potatoes, nuts and beans), sorghum, and millet (Kuper, 1963, p.34; Marwick, 1966, p.158). Cattle provided milk but, because of their ritual significance, were sparingly used for meat. Supplementary livestock included sheep, goats, dogs and chickens. Productive activity was marked by specific divisions of labour, and was carried out in an unchanging seasonal rhythm. Men's work included ploughing fields, milking cattle (a task taboo to women), sewing skins and cutting shields, hunting, and animal husbandry. Women and children carried out work of a continuous nature, including planting, manuring and weeding fields, carrying water, cooking, grinding grain, child rearing, and maintaining the homestead (Kuper, 1947, p.35; 1963, p.45).

Agricultural practices were essentially conservative, involving limited surplus production, in a manner consistent with major themes in subsistence and peasant economy (Allen, 1949, 1965; Chayanov, 1966; Foster, 1965, 1973; Mamdani, 1972; Sahlins, 1972; Scott, 1976). Kuper described this system of production as: '... living on a precarious subsistence level ... [like] all primitive peasants' (1947, p.36), though her data reveal that the Swazi diet was not meagre, and was quite diverse.

The traditional female dominated homestead economy was geared to the systematic occupation of new land in a militarist state. By accepting the judicial authority of the chief and the king, and the membership of an age regiment, which demonstrated a willingness to go to war in times of need, men held an unalienable right to receive land. Female labour freed men for war, and the levirate and sororal polygyny ensured the continuance of the patriline, and continuing rights to land despite war deaths. The loyalty of men was further encouraged by the institutions of *lobola* (marriage cattle) and *sisa* (cattle lending arrangements), which ensured that cattle were a rare and prized commodity. Cattle could be won as war booty, and then be used either as marriage cattle or, through lending arrangements, to strengthen bonds of obligation between kinsmen.

While the political and judicial authority of the king was absolute, the state was not economically dependent on the agricultural resources of the homesteads, but rather on the immediate availability of warriors. This allowed the Swazi to retain their independence throughout the turbulent period of the eighteenth century, in an environment of tension and shifting alliances. Subsequently, when men began moving to work in the South African goldmines in the late 1800s, the female dominated homestead economy supported male migrant labour as it had supported militarism.

Contemporary Swazi society and the Real Farmers

Contemporary Swazi society differs in numerous ways from traditional Swazi society. The most obvious differences are those of the division of land and of infrastructure (see Figure 1.2). Swaziland is now divided into areas of Swazi Nation Land and Individual Tenure Land (see Figure 1.3). Swazi Nation Land are those areas where land is used by rights of usufruct, under the political control of a chief. Individual Tenure Land is land governed by individual tenure. The decisive difference is that individual tenure embraces rights to sell land; rights which do not and cannot exist under traditional tenure, although rights to sell modern brick buildings on land held under such tenure, do exist, by agreement with one's chief.

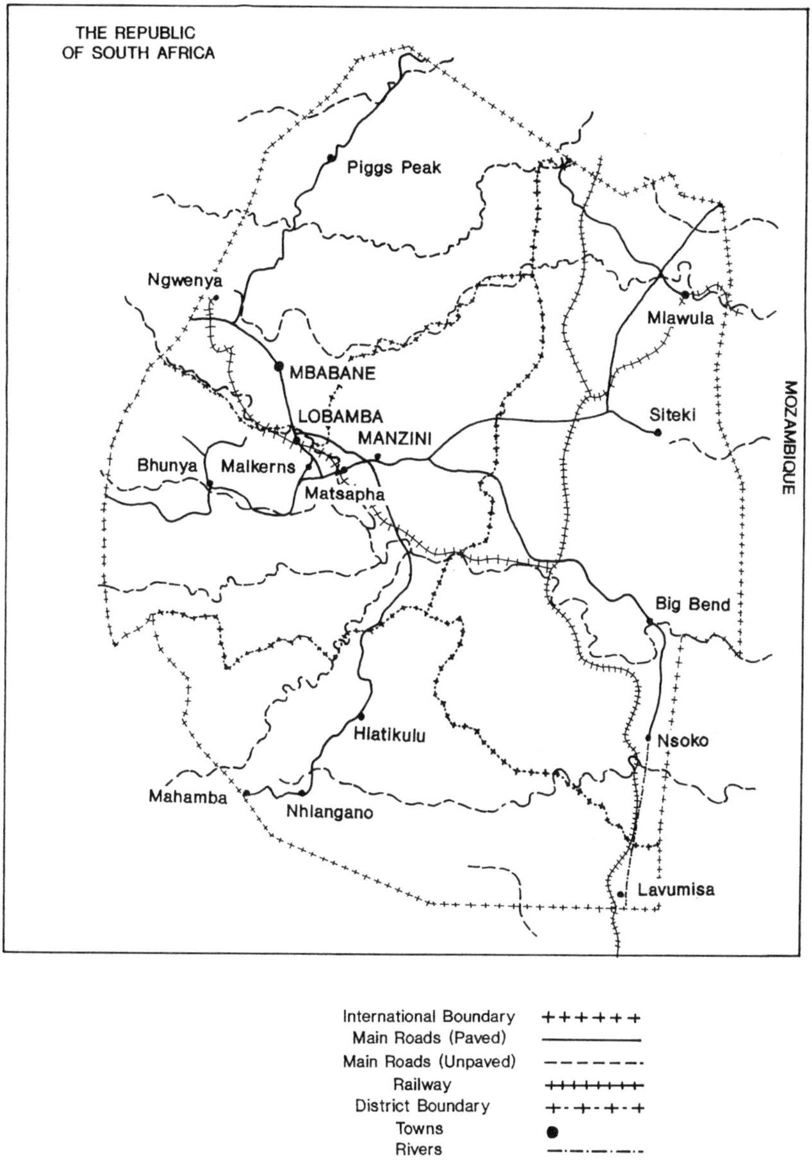

Figure 1.2 Swaziland

Figure 1.3 Land Tenure in Swaziland

Swaziland has several well developed urban centres, (Mbabane and Manzini), an extensive road and transport system, telecommunications, light industry, and highly developed agri-business operations. All these infrastructural changes have drawn many Swazi to the towns in search of job opportunities.

In recent years there has been a dramatic upsurge of urban, peri-urban and industrial development. During the 1970s the Ezulweni Valley, the Malkerns area, Manzini and Mbabane grew rapidly. For example, the squatter settlement of Nkwalini-Mahwalala, outside Mbabane, completely outstripped government predictions of growth, developing from a small squatter settlement in the early 1970s to a considerable township of 14,000 people in 1983 (Testerink and Russell, 1983a, pp.1, 6-7, 39).

A further example of a rapid growth area is that between Lozitha and Manzini. This area is part of the corridor following the roads from Mbabane to Manzini. A study of aerial photographs from 1959-1984, covering an area of 46 square kilometres, shows that in 1959, this area was characterized by open pasture, settlement areas comprising homesteads and maize fields, three farm houses for foreign farmers, the Kwaluseni Post Office, a single main road, many tracks, a primary school and the Matsapha National School (Dlamini, 1985, pp.1-9). By 1971 there were, additionally, five square kilometres of industrial area, a hydro-electric scheme on the river, two main roads, a prison, a medical centre, an army base, a police training college, an airport and associated housing, and greatly expanded suburbs emanating from Manzini. Much squatter housing had sprung up along the periphery of the industrial area, while along the main roads there was a dramatic increase in homestead numbers. A Rural Development Area Centre had been built along with its associated permanent buildings. A consequent drastic reduction in pasture land was also noted (Dlamini, 1985, pp.9-12). By 1985 the airport had been extended. The industrial area in Matsapha had grown in scale, though not in total area. High density flats, a supermarket, shopping complex with supermarket and fuel stations, and the University of Swaziland had been built nearby at Kwaluseni. There was a noticeable increase in homesteads servicing Matsapha and Kwaluseni and pasture land had been further drastically reduced to less than ten square kilometres (Dlamini, 1985, pp.16-17).

Despite the changes that have occurred, many elements of traditional social organization and subsistence production are evident in the contemporary homestead. For example, many farmers still favour agricultural techniques that give them a guaranteed subsistence yield at the expense of surplus production (Black-Michaud, 1981; Sibisi, 1982).

Within the government, the King has functioned as an absolute monarch since 1973. While royal power has been sustained by the King's control over the apparatus of the state, the power of the aristocracy at the local level, which was previously based on the allocation of land, has diminished with the increasing scarcity of arable land. The degree to which the King controls the apparatus of state is illustrated in Figure 1.4.

Contemporary Swazi society is more differentiated than it was in the past. While aristocrats and commoners are still evident, a further three social categories have emerged. An urban elite is prominent in the cities, and plays a significant role in the government, bureaucracy, and business sectors. This urban elite has emerged from among the former indigenous colonial middle class. Bi-lingual and educated, its members are men and women who responded to the many employment opportunities after independence in 1968. Wage and salary earners are another category of people who have become significant over the last twenty five years. The development of urban industries, service industries, and large agri-business conglomerates, has provided many opportunities for salary and wage work, and has drawn many people to the city. The fifth category is the *Bantfu Mbamba* (lit., 'Real People'). The Real People are a rural based category of people who are recognized by their contemporaries as having more than average drive and ambition. Mbamba translates literally as 'actual' or 'real' (Rycroft, 1981, p.61). However, English speaking informants give the word 'real' an emphasis which suggests that Real People have additional substance compared to others. When informants indicate a Real Person, they usually comment on the qualities that are indicative of that designation. For example, the Chief described to me a subject of his, whom I knew from others to be a Real Person, in the following terms: 'He is strong, he has been here a long time, and he is reliable, he is someone you can ask because he knows the way'.

In this particular instance the Chief was telling me that the man in question was noted for his dependability and his impartiality, and for those reasons he was given positions of responsibility on the Chief's councils. In the Chief's opinion it is essential that his other subjects be able to trust the men the Chief keeps around him.

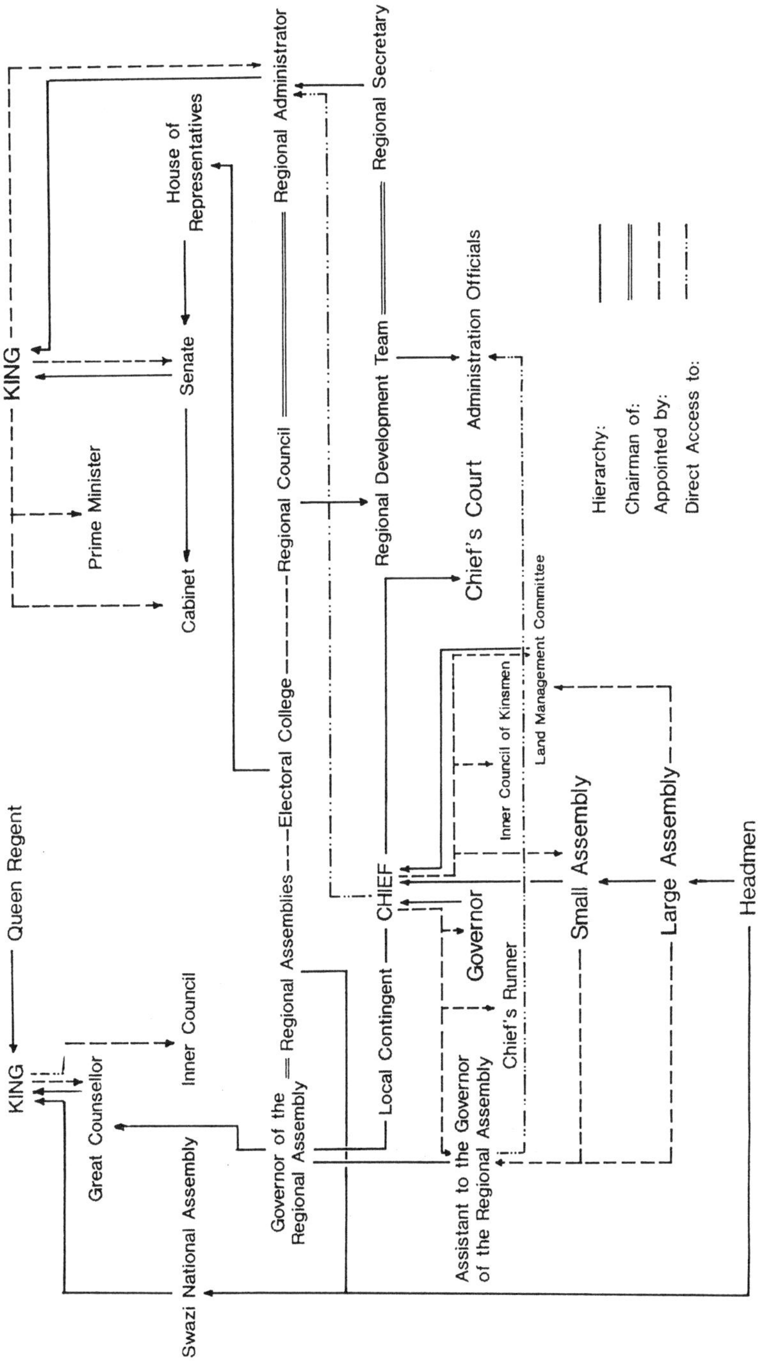

Figure 1.4 Swazi Government Structure

The Real People are those who are 'doing well' in Ngwempisi. They are people who have significant social and/or political profiles. They are not restricted to any particular occupation or social group. As a category of people, they are as diverse as the Chief, a wealthy supermarket owner, a subsistence farmer with a flair for organizing school building programmes, and a man who makes his living running a small cartage business. Real People are also farmers, labourers, and employees of government agencies. Real People are distinguished by their commitment to living in Ngwempisi, and they invest much time and energy in developing their local social and political profiles. As a category, the Real People incorporate a variety of sub-categories; the local government personnel, the chiefs and their headmen, and the *Balimi Mbamba* (Real Farmers) who are the subject of this book.

The Real Farmers are people, male or female, who have a primary commitment to commercial farming on Swazi Nation Land [5]. Real Farmers are distinguished, both by their long term goal of becoming economically self-sufficient, and by the degree to which they are involved in local politics and society. The Real Farmers are a loose grouping defined more by individual, rather than collective, identities. Despite this, the Real Farmers are readily identified by their fellows, and therefore have a publicly accepted profile. The Real Farmers are *balimi* (farmers). Farmers however, are not necessarily commercial producers, although the term does refer specifically to people involved with crop production, as opposed to commercial animal husbandry carried out by *bafuyi* (animal farmers; from *timfuyi* 'domestic animals').

A further qualification must be made with the use of the term headmen which is sometimes used to refer to Real People. Since Real Farmers are also Real People, the term headman can sometimes be used for Real Farmers. People tend to use the term headman when referring to the conservative rural wage earners rather than Real Farmers, but in general, the term can be used to describe people who are respected in the chiefdom, as they are in their own homesteads. Such people are described as dignified, approachable, trustworthy, and successful. While the Real Farmers may fit all these classifications, they are, above all, committed to the long term goal of successfully making their primary income through agriculture, on Swazi Nation Land.

Despite the emphasis on men as Real People and Real Farmers, women are significant members of both categories. Male Real Farmers are more conspicuous than women because men have always worked away from their homes and so those who stay to farm are immediately noticeable. Also, women have always dominated rural agricultural life, and they are always evident in the countryside. Establishing women's motivations for their agricultural work is a more difficult process, since women do not exercise such obvious career choices as men.

The emergence of women among the ranks of the Real Farmers reflects a long term interest that many women have in commercial ventures. Women have long dominated the markets in Swaziland, and more generally in Southern Africa. This situation evolved out of processes of labour migration, which took men away from their homesteads and isolated them from local commercial opportunities. Also, women have always had primary responsibility for their own children. This latter fact has limited the freedom of women to pursue work in the open labour market, and has also required that they maintain a constant and reliable source of income. As a result, in rural areas many women became involved in manufacturing handicrafts for the Swazi and South African tourist markets. Among Real People in Ngwempisi, several women are prominent as managers of enterprises of this kind. The attraction of commercial farming for enterprising women is that they can use farming skills they already have to develop a strong local market, rather than enter a risk prone external market, such as that of handicrafts.

In Ngwempisi, the Real Farmers are concentrated at Ntimane. It is difficult to be precise about exactly who is a Real Farmer since many farmers are secretive about their financial status. This prevented me from establishing exactly the degree to which some may successfully earn their primary income through commercial farming. Such secrecy also makes it difficult to determine the long term intentions of farmers. At Motane, a small irrigation scheme near Ntimane, I identified one farmer who is definitely a Real Farmer, on the basis of his successful achievement of his goal of commercial farming and his high local profile. In that same scheme, several other members are successful farmers, but I could not ascertain their financial status.

Three other farmers, who are not members of irrigation schemes, are also Real Farmers. One grows maize at his homestead. He uses chemical fertilisers and hybrid grains to achieve a substantial surplus yield every year. The other two farmers separately irrigate small pieces of land near the river, where they grow vegetables for sale to a local clientele. In both cases they successfully earn their entire income from their farming operations, and have significant social profiles.

The largest concentration of Real Farmers is at Ntimane, where I identified twenty five people, ten women and fifteen men, whom I consider to be Real Farmers according to the above description, and which I later confirmed by asking other people to identify the Real Farmers they knew. Of these Real Farmers, approximately eighteen receive their primary income from farming at Ntimane. The other seven earn more money from enterprises other than farming, but are highly committed to their commercial operations at Ntimane.

Literature and explanations

The Real Farmers are now a significant part of rural society. Despite this, they are largely unacknowledged in the literature pertaining to contemporary Swazi society. The current view of Swazi society, held by non-Swazi, is based on a model developed at a time when social structure still largely conformed to that of the pre-colonial period. In essence, anthropological and historical views of Swazi society promoted a model of the rural homestead as a discrete economic entity, with a strong subsistence base, linked to the state by the king's control over land. The most noticeable proponent of this form of description was the anthropologist Hilda Kuper [nee Beemer] (Beemer, 1935, 1937; Kuper, 1941, 1944, 1945, 1946, 1947, 1950, 1963, 1964, 1968, 1971, 1972a, 1972b, 1973, 1975, 1978). Kuper's strong structural functionalist anthropological descriptions are still used by many researchers as an ethnographic data base. While Kuper was the most prominent of the writers from the period, other descriptions substantiate her work. Kuper, Marwick (1966), Cook (1930), and Engelbrecht (1930), all envisaged social change as constituting a move away from a traditional ideal in a society with a fixed hierarchy. Furthermore, while acknowledging a degree of contact between rural Swazi and the European dominated market economy of South Africa, these commentators did not pursue the implications of such contact on social organization in Swaziland, although Kuper (1947, p.68) did acknowledge the possibility of limited social mobility.

The structural functionalist view of contemporary Swazi society continues to be propagated. For example, popular publications aimed at the tourist market (Schwager, 1984), and histories aimed at Swazi school children (Matsebula, 1980), emphasize a juxtaposition between a 'traditional sector' and a 'modern sector'. Media representations, particularly from outside Swaziland, stress the traditional aspect of Swazi life (New African, October 1985). For example, in a BBC documentary, 'Lion of Swaziland', part of the reputable series 'The World About Us', the narrator stated:

> With the King's encouragement and example, the traditional life of Swaziland goes on in many respects as it always has. The villages and towns are few; the Swazi preferring to live in homesteads scattered about the plains (Cartwright, 1982).

This is a view at odds with much of the contemporary Swazi media, as represented by Swazi journalists, who perceive no ambiguities in the way Swazi live their lives (*cf.* Swazi Life). This lack of perceived ambiguity is the dominant view of life shared by most Swazi. They have an awareness of tradition but do not see tradition as inhibiting the way they live. Also, they do not perceive themselves to live in a rigidly ordered hierarchy without the possibility of upward social mobility. Furthermore, they do not recognize any substantial distinction within Swazi society in strong dualist terms.

The dualist view of Swazi society has continued to permeate critical analyses of contemporary Swazi society, despite the views of the general population. For example, in the area of rural development, the entire Rural Development Areas Programme was formulated on the basis of a belief that Swaziland had a 'dual' economy. This perspective assumes a particular view of the Swazi rural economy, and of the relationship between the rural community and the state. This relationship has been described as one in which social change occurs as a unilinear progression, away from traditional practices, to a model of a Western industrialized democratic society, in which transition is basically a technical problem in a pre-rational and pre-industrial society (Higgott, 1983, p.18; Higgott and Robison, 1986b, p.18; Nafziger, 1979; O'Brien, 1979, p.55; Randell and Theobald, 1985, p.35,67; Robertson, 1984, p.28; Webster, 1984, pp.41-63).

This view of Swazi society is most pervasive in the National Development Plans (Government of Swaziland, 1978a, 1985a), and a series of subsidiary publications (Government of Swaziland, 1976b, 1978b, 1979b, 1983, 1983b, 1985b). In these publications a traditional/modernity dichotomy is accentuated, and tradition is proposed as an obstacle to development. The Swazi Government is supported in this view, both financially and conceptually, by many foreign governments. The bulk of the research into development and change is financed, either directly or indirectly, through aid packages, and the resulting publications reflect that aid based research, though individual researchers do bring their own ideas to bear.

For the most part, these aid funded publications are similar, expressed in technocratic terms, with a strong reliance on statistical data, and without regard for the influence of social processes on economic behaviour. Particularly noticeable in this category are the publications of Cohen, published under the auspices of the United States Agency for International Development (USAID) (Cohen, 1983, 1984a, 1984b, 1984c, 1984d, 1985a, 1985b, 1985c; Cohen and Green, 1984). Other USAID publications are similarly premised on principles that emphasize formal approaches to economic theory and promote a sharply dualistic analysis (USAID, 1978a, 1978b, 1979, 1985). The Commonwealth Development Corporation, which has a long term involvement in Swaziland at Vuvulane, provides reports that are merely statements of accounts and venture no comment on the actual operations of the corporation in sociological terms (1976, 1983, 1984a).[6] Any efforts to provide social data about Vuvulane are the work of individuals involved in management, rather than part of corporation policy (Commonwealth Development Corporation, 1984b; Tuckett, 1975).

Some researchers and commentators have been certain of the dualist nature of Swazi society. For example, Poulson (1983) suggested a model for community planning wherein:

> People are now expected to revise or review traditional solutions in the light of modern knowledge and national requirements, and to arrive at solutions that are realistic for balanced local and national development (Poulson, 1983, p.278).

The majority of researchers arguing from this perspective, like Poulson, tend to provide research that is abundant in statistical data, but from which few conclusions are drawn (*cf.* Vost, 1979).[7] This trend began with the Random Sample Survey of July 1960 (Holleman, 1964a), which, although a wide ranging, rigourous, and illuminating series of studies (*cf.* Banford, 1964; Cruise and Holleman, 1964; Daniel, 1964; Holleman, 1964b, 1964c, 1964d; Hughes, 1964a, 1964b, 1964c; Humphreys and McClelland, 1964; Jooste, 1964), set a precedent which continued under the restraints of the post-independence aid based economy. For example, much of the more recent work of this kind has been carried out under the auspices of the Social Sciences Research Unit (SSRU) at the University of Swaziland. The Social Sciences Research Unit sponsors an eclectic group of researchers (*cf.* Capetta, 1983; De Vletter, 1983b; Guma, 1983; Huppert, 1983; Mercy, 1983; Russell, 1985b; Russell and Ntshingila, 1984; Testerink, 1984; Testerink, Funnell and Freund, 1985). These researchers are often carrying out work for aid agencies and the Swazi Government, ([UNICEF] Armstrong and Russell, 1985; [the Swazi Government housing branch] Testerink and Russell, 1983, [FAO] Funnell, 1982a, 1982b; Russell 1985b; [USAID] Armstrong, 1985), and more rigourous theoretical analyses in their research tend to be subordinated to general governmental concerns with solving problems associated with the implementation of rural development in Swaziland.

The prominence of research predicated on a dualistic view of Swazi society also forms the basis of much general research (De Vletter, 1982; Green, 1986; Guma and Simelane, 1982; Kliest, 1982; Low, 1986; Magagula, 1978, 1980, 1981, 1982a, 1982b; Ntshingila, 1984; Testerink-Maas, 1985). This situation is partly an artefact of the manner in which funding is tied to pre-determined analytical frameworks.

The view of Swazi society as dualistic is largely supported by another group of commentators who are, nevertheless, critical of how social change occurs in Swaziland, and also of the nature of the relationship between the community and the state. The bulk of these criticisms identifies the relationship between Swaziland and the world economic system as largely determining the nature of contemporary Swazi society. These theories fall roughly into three groups. First, there is the view that structures inherited from colonialism blocked capitalist development in Swaziland. Second, it was posited that semi-peripheral states may exist in a dependent capitalist relationship between the core and the periphery. Finally, there was a micro view of local structures, in which pre-capitalist and capitalist 'modes of production' interact in varying social formations (Higgott, 1983, pp.48, 58-62; Higgott and Robison, 1986b, pp.30, 64; Randell and Theobald, 1985, pp.132, 138).

Daniel (1979a, 1979b, 1981, 1984, 1986; Cohen and Daniel, 1981; Daniel and Stephen, 1986) argues that colonial structures blocked capitalist development in Swaziland. He further suggests that, in the process, colonialism imbued subject peoples with an alien personality that detracted from their ethnicity (1979a, 1979b, 1981). Daniel (1984) later modified this argument, by proposing that vested interests existed among the Swazi elite, which encouraged a dependent relationship with South Africa. This perspective is complemented by recent writings from the Centre of African Studies at Eduardo Mondlane University, Mozambique. These writings assert that Swazi contact with international capital is reinforced through their client status with South Africa, and maintained by South African de-stabilisation (Davies, O'Meara and Dlamini, 1985). This is a view supported in Swaziland by Stephen (1986a, 1986b; Daniel and Stephen, 1986), and Neocosmos (1987a, 1987b, 1987c, 1987d).

Levin (1981, 1984, 1985, 1987a, 1987b), on the other hand, while concerned with the relationship between Swaziland and the world economic system, uses micro-economic data. He differentiates between pre-capitalist and capitalist modes of production in contemporary society. He emphasizes class formation rather than class conflict, and presents evidence for uneven development rather than blocked development. Levin's analyses of the nature of contemporary society are more rigourous in terms of local data, and provide valuable insights into development and change. Levin's data and analyses come closest to identifying the emergence in Swazi society of categories of people who play an active part in social change. Standing (1987) and McFadden (1986), support Levin's emphases on local social categories, albeit without resort to similarly rigourous analyses of local empirical data. Standing is highly critical of the way many assumptions about the nature of contemporary Swazi society remain unchallenged (1987, p.127). By arguing that a dearth of empirical data disguises the changing economic role of women, Standing concludes that growing differentiation between women in rural homesteads can be related to uneven development in Swaziland. McFadden (1986) argues that blocked development further reduces female status in wage labour.

Similarly, Packard (1986), Simelane (1986) and Vilane (1986) provide insights into local social processes in Swaziland. Packard (1986) argues that processes of development reinforce unequal access to land, and perpetuate disruptive labour patterns in a manner which facilitates outbreaks of malaria and malnutrition. Vilane (1986) and Simelane (1986) suggest that the nature of development has encouraged royal hegemony at the expense of the rights of wage labourers, leading to a society that is increasingly materially differentiated.

The analyses of Swazi society that have stressed macro-economic data, at the expense of local empirical data collection, have added to the general understanding of observers as to the nature of contemporary Swazi society, but have had a limited impact on planning for development, or understanding social change in Swaziland. Whilst these researchers have been the most trenchant critics of the rationale buttressing rural development, they have also tended to reiterate a similar dualist view of Swazi society.

The bulk of research material on Swazi society, development and change, supports neither the polarised views on the nature of the rural economy and the degree of its integration with the state and international capital, nor the earlier functionalist perspectives of Kuper and Marwick. As a result, much research carried out in Swaziland employs an interesting blend of analytical styles. This research is focused more on data indicative of local class structures related to, or emerging because of, social change.

Representative of this category are writers such as MacMillan (1986), who deals at length with the flexibility of tradition, and the development of class divisions in Swazi society over a long period of time. Matthews (1987) criticises the idea that the homestead is homogeneous, and provides interesting insights into the development of indigenous class solidarity as opposed to national identity. Youe (1986) provides a precise historical account of the evolution of land claims in Swaziland. Neocosmos and Testerink (1985), in an advocacy of the need for more empirical data, propose methods of examining contract farming operations. Russell (1986), examining the role of women in development, suggests a cautious analysis in line with her earlier work (1983a), in which she examines processes of production in the rural homestead, and suggests that the homestead provides a buffer against the emergence of 'grosser inequalities' in Swazi society (1983a, p.21).

Low (1983), moving away from the formal approach he adopted in his earlier research (Low, 1982; Low, Kemp and Doran, 1980a, 1980b), calls for an examination of socio-economic constraints on development in the homestead. Low asserts that aid is tied to 'attractive production projections' (being attractive to the development planners), and perpetuates inadequate examination of socio-economic factors in the homestead (1983, p.160). Indirectly, he is therefore critical of the dominant view of a dual society. Particularly significant in this general category are writers such as Crush (1977, 1980a, 1980b), who attacks the view that social change is a technical matter. Crush reacts against the concept of a plural society, and argues strongly for an examination of the manner in which the colonial state responded to the interests of the Swazi royal family, and not solely in the interests of foreign capital. In this, he is supported by historians such as Booth (1983, 1986) and Bonner (1983), both of whom have written significant books on Swaziland from a historical perspective. These researchers provide imaginative and original insights into the adaptive mechanisms of the Swazi homestead in adverse economic and social conditions.

A variety of researchers have moved more fully into forms of analysis involving a greater stress on examining the dynamics of the homestead in order to ascertain the active way rural Swazi relate to external wage opportunities. Russell accepts change as a social reality but emphasizes the ongoing integrity of the homestead. She argues not that the homestead has become part of the capitalist mode of production, but rather that capitalism has become part of rural domestic production (1983a, p.319). Russell states that: '... the rural homestead is deeply and apparently increasingly engaged in off farm employment for wages, and cannot be understood except as an integral part of the total [Swazi] economy (1983a, p.317)'.

Russell denies any economic or social duality, and suggests instead that involvement in rural wage activities represents '... deliberate decisions to disperse homestead labour in a more profitable way' (1983a, p.328). In Russell's view, the homestead is the rural base for the extended family, provides security for the aged and, through selective entry into the cash economy, is based on a subsistence ethos (Russell, Mbatha and Sithole, 1982; Russell, 1983a, 1983b, 1983c, 1984, 1985a, 1985b).

Ngubane (1983) (*cf.* Sibisi, 1979, 1980, 1982)[8] supports Russell's contention that the homestead is highly adaptive and selective in exploiting cash earning opportunities. Ngubane further suggests that cultural integration is not only part of Swazi cultural practice, but common to most African cultures (1983, p.3). Russell's and Ngubane's arguments are supported to some degree by a number of researchers. For example, De Vletter (1983a, 1983b) and Low (1982, 1983; Low, Kemp and Doran, 1980a, 1980b), working within a formal economic framework, view cash activities in the homestead as highly selective and logistically sophisticated.

The bulk of research on the nature of Swazi society has either tended to over emphasize economic and political duality, or has failed to come to terms with social change and the emergence of local social structures in Swaziland. By using anthropological data, which are now over fifty years old, for the social framework of contemporary enquiry, researchers have largely failed to account for either the rapid growth or the highly economically and politically integrated nature of rural society in Swaziland.

The emergence of the Real People is one major indicator of the degree to which political and economic integration is a reality in Swaziland. Furthermore, when Swazi talked of tradition and modernity, no sense of disjunction between the two concepts existed. In the minds of people the concepts and the material items exist as realities, but are not in any way mutually exclusive. However, rural development policy since 1968 was formulated on the basis of a dual sector model of contemporary Swazi society. Expatriate planners seem to have interpreted noticeable material differences between urban and rural areas as an affirmation of a dual society, and accepted this evidence as an appropriate view of Swazi society. It is more difficult to conclude that the Swazi Government, especially King Sobhuza II, perceived Swaziland as a dual society in this way. Despite this, the Rural Development Areas Programme went ahead and all the indications are that it has not achieved its set goals. This is not to deny that much has been achieved because of the programme.

There is little concern among farmers and programme officials over the demise of the programme, or the difficulty in achieving set goals. To the contrary, the programme represented many things to many people, and it gave the Real Farmers their best opportunity to emerge as they have. However, the logic of opportunity for the Real Farmers, while far from precise, can be simply expressed. For example, a Real Farmer responded to the news that Ntimane was going to buy a trailer, despite the fact that the Ngwempisi Rural Development Area made this equipment available, by saying:

> The news is very good news, but we must all be grateful to the people at the Ngwempisi Rural Development Area because they have taught us the importance of being self sufficient in this

equipment. Because they have refused to let us use the equipment, they have taught us the importance of buying our own.

The Real Farmer, in making this statement, demonstrates that he perceives opportunity even in adversity. One benefit of the programme can thus be interpreted as the provision of crucial lessons in self reliance for the farmers.

The Real Farmers have emerged in Ngwempisi because of, and also despite, the Rural Development Areas Programme. They have both used and rejected the programme, using the available resources, but remaining aloof from the organization. They have emerged as the significant economic force in the area, and are surpassing the ability of the Rural Development Areas Programme to do anything for them.

The relative success of the programme rests on how its success or otherwise may be defined, or on what basis it may be analysed. While goals have not been achieved in terms of infrastructure and production, rural development has occurred, and is comprehensive. The programme planners did not anticipate the emergence of people such as the Real Farmers, nor did they anticipate their power and influence, or the way that they would reinforce centralized royal power in the Swazi state.

Land scarcity and rural development

The advent of British control in Swaziland, after the Second Boer War in 1902, radically reduced Swazi access to land. The Swazi were prevented from seizing more land, and their political boundaries were defined. The most significant decision the colonial authorities made was to convert the concessions held by European settlers into freehold land. This process alienated the Swazi from sixty per cent of the land. By 1920, the Swazi population was struggling to cultivate the thirty two delineated 'Native Areas', on what proved to be the poorest land in Swaziland (Crush, 1980a, 1980b). British colonial development policies, in response to over crowding on Swazi Nation Land, were implemented from 1944 to 1965 under the Native Lands Scheme (Funnell, 1984, p.1), but were sporadic and desultory, favouring the wholesale utilisation of Swazi resources by farmers and business. After World War II, this activity expanded into forestry, agri-business, mining, light industry and tourism (Barker, 1965), with commercial opportunities for Swazi confined to wage work in large scale, labour intensive industries on Individual Tenure Land. The exception was a small holder scheme, the Vuvulane Irrigated Farms, which promoted sugarcane growing to supply the European market (Commonwealth Development Corporation, 1984b; Tuckett, 1975).

In 1968, the Swazi claimed compensation for the alienation of land, and also compensation for the 500,000 acres the British had sold to fund the colonial administration. The British recognized the legitimacy of Swazi claims, but rejected the claim outright and suggested land development projects instead (Kuper, 1978, p.298). These projects, funded with 1.5 million pounds and based on the previous colonial scheme, would target specific areas of Swazi Nation Land for intensive development (Davis, 1980, pp.5, 27). Under the terms of the aid package, the land was not to revert to traditional patterns of use, and a 'positive' development effort was to be made (Funnell, 1982b, pp.274-275).

The Rural Development Areas Programme had two major objectives: to involve the majority of people in development in order to improve the quality of their lives, and to reduce Swazi dependence on external finance. The primary emphasis of development was to be on the intensification of agricultural production on Swazi Nation Land, from traditional subsistence agriculture to semi-commercial and commercial farming, in an attempt to create more employment opportunities, and to raise incomes in rural areas (Government of Swaziland, 1978b, pp.xi, 23). A major assumption was that farmers would readily adopt a more commercial attitude to farming, so this advance would be achieved through an integrated package:

> ... for strengthening extension and credit services, livestock development, land development and conservation, incremental farm inputs and technical assistance (Government of Swaziland, 1985, p.146).

The Rural Development Areas Programme

The Rural Development Areas Programme began officially in 1971. By 1976 there were four Rural Development Areas, including eleven per cent of the rural population and totalling 71,000 hectares (Funnell, 1982b, p.275). In

1978, after the Ngwempisi Rural Development Area was established, funding from the United Kingdom ceased (Davis, 1980, p.31), but was taken over by a consortium of aid organizations, including the World Bank, the African Development Bank and the European Development Fund. The dominant contributor was the USAID, through the Rural Development Areas-Infrastructure Support (RDA-IS) Project, which aimed to bolster the programme to the extent of USD 14,944,000 (USAID, 1978a, p.1; 1985, p.5), in the belief that such aid was a valid and cost effective method of developing Swaziland's 'sharply dualistic' economy (USAID, 1978b, pp.1-2). The RDA-IS project ceased in 1985 (USAID, 1985, pp.1-7), but by 1986, and from a variety of sources, the US had contributed an estimated USD 76,009,000 without result, supporting the contention that USAID was '... throwing money at the problem' (Funnell, 1982a, p.1).

In 1982, after Sobhuza's death, the government continued to endorse the Rural Development Areas Programme, but clear evidence emerged of a diversion of effort by the government, into the funding of large scale agri-businesses on Swazi Nation Land and Individual Tenure Land. The EEC, through the LOME agreements, provided access to European markets for Swazi produce, and made substantial aid and trade arrangements (Times of Swaziland, 29/5/85).[9]

Further arrangements involved the Southern African Development Co-ordination Conference, the European Development Bank, the Canadian Government, the Republic of China (Taiwan), the United Nations and many other smaller agencies which still have programmes in Swaziland.

The Rural Development Areas Programme had lost all impetus by 1985, when the Fourth National Development Plan (1983/84-1987/88) was published. In that plan, little reference was made to small scale development on Swazi Nation Land, other than a dismissive statement: 'Production on traditionally held land [Swazi Nation Land] has been less than expected' (Government of Swaziland, 1985, pp.i, 139). However, an explicit conclusion was drawn that:

> ... before the problem [poor production] was defined, the solution was institutionalised. The effects of the [Rural Development Areas] programme are that a transfer of resources from the modern to the traditional sector has taken place, without the productive base to finance and maintain that transfer

having been established (Government of Swaziland, 1985, p.166).

The difficulties experienced by the Rural Development Areas Programme in achieving set goals can be measured in terms of a lack of self sufficiency in maize production, and a general increase in imports of maize. Maize production actually dropped by 5.3 per cent, partially replaced by a growing public preference for vegetables. Production was also affected by increased imports to the point where imported maize matched local maize production by volume. Cotton farming over the period showed no significant increase in yield, and there was also no difference in production between development areas and other areas. Tobacco production actually declined over the period (Government of Swaziland, 1985, pp.147-152, 166).

The move away from further small scale development on Swazi Nation Land, and growing evidence of government concerns with the private sector, were reflected in the general economic indices, and the nature of government spending up to 1985. Recurrent spending in the agricultural sector dropped forty-four per cent, while increasing by fifty-eight per cent in industry and mining, and by eighty-nine per cent in water and sewerage. Companies received tax incentives, while spending on infrastructure increased at the expense of schools and health clinics (Cohen, 1985b, pp.1-6).

By 1983, the external public and publicly guaranteed debt doubled to thirty-three per cent of Gross Domestic Product [from E43.9 million in 1978/79 to E187.5 million in 1982/83]. During the year 1983/84, the debt rose to E232 million, and then in the 1984/85 year, partly because of the fall of the South African Rand, the debt rose to E326 million (Swazi News, 9/11/85). From 1979 to 1983, domestic demand outgrew domestic supply, resulting in increased funding from external sources. The main demand was in the area of recurrent spending which, as a share of overall expenditure and net lending, grew from 41.1 per cent in 1978 to 62.7 per cent in 1983. Personnel costs comprised over fifty per cent of total recurrent expenditure [55.4 per cent in 1978 to 53.7 per cent in 1983], because of large increases in informal employment by ministries, and high demands for support infrastructure (Government of Swaziland, 1985, pp.110-111). The blowout in recurrent expenditure was unanticipated. The proposed budget in the Third National Development Plan, 1978/79-1982/83, suggested that recurrent expenditure would move from E55 million in 1978, to E69 million in 1983.

In fact, the increase was from E59.2 million in 1978, to E124 million in 1983 (Government of Swaziland, 1978b, p.46; 1985, p.111).

By 1985, concerns among donor governments over trends in the Swazi economy resulted in a report from a visiting World Bank team, which proposed more effective monitoring and control of parastatals (Cohen, 1985b). Real growth until 1990 was forecast at 0.8 per cent (Swazi Observer, 6/6/85). The government authority responsible for the Rural Development Areas Programme was unable to offer any answers as to why small scale farming on Swazi Nation Land had failed to produce any substantial surplus, other than to comment that it:

> ... was assumed that SNL farmers would readily adopt a more commercial attitude to farming, if they were given inputs and advice. Large increases in the production of maize and particularly commercial crops were foreseen (Government of Swaziland, 1985, p.147).

Current rural development policy

Despite obvious streaming of funds to large scale industries, development on Swazi Nation Land still poses some dilemmas for the government. This is reflected in on going but desultory efforts to promote cash cropping. Official policy initiatives now concentrate on supporting self help projects. This support has taken the form of a proposed central marketing project costing USD 6.2 million (Swazi News, 19/4/86), and the less than popular *Ekhaya* ('go home!')[10] project of the then Prime Minister, Prince Bhekimpi, in which it was proposed to force the unemployed out of the city and back to the land, where they could produce food for the nation (Swazi Observer, 30/8/85; Times of Swaziland, 13/9/85, 30/8/85). This project was to be supported by a donation of 22 tractors to various chiefdoms, in a deal sponsored by Swazi Ford. This soon soured (Swazi Observer, 1/11/85; Times of Swaziland, 9/11/85, 18/12/85).[11]

International support for Swazi Government initiatives was reflected by USAID which reiterated its commitment to aiding self help projects (Times of Swaziland, 21/3/86), as did the Unitarian Services Committee of Canada (Swazi Observer, 23/4/86). Other organizations have also supported small self help schemes. The American organization, the National Council of

Negro Women, for example, has funded several small self help piggery units (Swazi Observer, 19/8/85). Much work is being done with smaller schemes by various agencies, such as the Chinese Agricultural Mission (Taiwan), the Africa Cooperative Trust, the Swazi Savings and Development Bank, and the Bonginhosi Tractor Company.

The 'go home' programme quickly declined into inactivity and Bhekimpi soon fell from power. The rhetoric of the new King reflected a continuity in emphasis on government sponsorship of private sector growth. Speaking at the opening of a new factory in Matsapha industrial area, he made a strong commitment to maintaining 'cordial industrial relations' in order to attract increased foreign investment and thus create more job opportunities in Swaziland (Swazi Observer, 21/9/87).

The government no longer tries to provide cheap services to farmers. For example, the tractor pool, once a cornerstone of the Rural Development Areas Programme, has lifted hire charges by between 30 per cent and 300 per cent, to rates well beyond those of private contractors (Times of Swaziland, 1/10/87). The government also spends large sums of money, much of it foreign aid, on maintaining its infrastructure. The West German Government, through the West German International Development Agency, spent E15 million on a government hospital and equity participation in the new industrial development body, the Swazi Industrial Development Corporation (Swazi News, 10/10/87). The Swazi Government is spending large sums of money on parastatals with severe budgeting problems which are not easily monitored; Royal Swazi Airlines, the National Development Corporation of Swaziland, the Swaziland Electrical Board, Swaziland Railways, Swazibank, the Swaziland Royal Insurance Corporation, and the Small Enterprise Development Company (Times of Swaziland, 21/10/87).

Increasing government pre-occupation with the private and parastatal sectors, and the evident decrease in capital funding to the Rural Development Areas Programme, indicate that while recurrent spending is maintaining bureaucratic infrastructures in the programme, no new initiatives are being taken. The implied support by government leaders and aid agencies for self help schemes on Swazi Nation Land, is partly a recognition of the relative success of such schemes, but is also offered as a solution to the cessation of centralized planning for rural development on Swazi Nation Land through the Rural Development Areas Programme.

As the government de-emphasizes support for rural development programming in the informal sector, it is increasingly sponsoring private industry. For example, by 1984 the Swaziland sugar industry employed approximately 13,000 Swazi, and approximately fourteen per cent of the population depended on it for a living. The industry was based on a network of foreign corporations (mainly British and South African), Commonwealth Development Corporation and Swazi ownership, with the vast bulk of equity in foreign hands (Levin, 1984, p.1). For example, the Commonwealth Development Corporation, in partnership with the Swazi Government, was by 1986 extending its involvement in Swaziland, largely by providing finance to enlarge and diversify existing projects (Times of Swaziland, 23/5/86). Under the Lome Conventions, aid and trade had by 1985 realized E220 million, of which E30 million was allocated directly through the period 1976-1985. From 1986 to 1990 this will be E50 million. These monies will take the form of duty free access to the European Economic Community for sugar, beef, tinned pineapple, pine furniture and other primary and manufactured goods, few of which are produced by small producers on Swazi Nation Land. By contrast, EEC support through the Lome Conventions for small scale producers on Swazi Nation Land amounted in 1986 to 5,500 tonnes of food aid, and E1.5 million for rural development (Swazi Observer, 1/6/85; Times of Swaziland, 29/5/85; Weekend Observer, 7/6/85).

The Simunye Sugar Company is diversifying into other areas of production such as cotton and beans, as a result of the down turn in the market for sugar in the early 1980s (Swazi Observer, 25/10/85). The Tibiyo Taka Ngwane corporation continues to grow through involvement with many large scale operations (Swazi Observer, 8/11/85) and, as a major partner with the Commonwealth Development Corporation, experienced a boom after an unexpected up swing in sugar and wood pulp prices during 1986. The Tibiyo Taka Ngwane corporation is also involved with the Mhlume Sugar Company, the Inyoni Yami Scheme (a citrus growing operation), as well as having a share in the Simunye Sugar operations (Times of Swaziland, 21/7/87).

Conclusion

The years since World War II have seen a plethora of development projects grow up on Swazi Nation Land, all aimed at developing that sector of the economy. The Real Farmers have emerged from within this process of change over a long period of time. They are a social category whose emergence has not been anticipated, and who are highly integrated with the local community, and ultimately with the Swazi state. The Real Farmers have emerged both because of, and despite, the Rural Development Areas Programme.

Notes

1. From the SiZulu for Mswati (Mswazi). Thus the people of Mswati were '*aba ka Mswazi*'. This term was then shortened to '*basonase*' and eventually to 'Swazi' (Jones, 1981, p.349).
2. 'The time of crushing' (lit., 'to make small'). This was a turbulent time in Southern Africa as the Zulu, under Shaka, and the Ndebele, under Mzilikazi, sought to expand their respective empires.
3. Swaziland has collections ranging from artifacts in the Oldowan and Acheulian traditions, to pottery and worked iron artifacts dated at AD 1000 (Harris, 1980, pp. 43-44; Hunter and Whitten, 1977, pp.32-44; Masson, 1982, pp. 2-3; Matsebula, 1980, pp. 1-2).
4. The British insisted that Sobhuza could not use the title of 'King' but must be called 'Paramount Chief'; instead, the people called him *Ngwenyama* Sobhuza II. *Ngwenyama* translates literally as 'lion' (Kuper, 1978, p.73; Matsebula, 1980, p.164).
5. Swaziland was divided into areas of Swazi Nation Land and Individual Tenure Land in 1907 as a result of the Land Partition Proclamation.
6. Commonwealth Development Corporation has a major involvement in Swaziland with The Vuvulane Irrigated Farms which, in 1986 for example, were experiencing a great deal of internal strife. Farmers were not happy for a variety of reasons and the international sugar market was not yielding a good return.
7. In presenting statistics on peri-natal mortality in rural Swaziland, Vost makes no attempt to relate peri-natal mortality to current social conditions, or to the logistical limitations implicit in pregnant women gaining access to facilities.
8. Ngubane originally published under the name of Sibisi prior to her marriage.

9. Under Lome One (1975-1980), three grants, totalling E13 million, were made to off set run downs in iron exports. As a result of Lome Two (1980-1985), two grants totalling E9 million, were made to off set run downs in cotton exports. Lome Three (1985-1990), resulted in grants of E36 million for National Development, though over time this figure may even be adjusted as high as E50 million (Times of Swaziland, 19/6/85, 29/5/85). Further subsidiary trade arrangements bring the total value of Lome, over the period 1975-1990, to E220 million (Weekend Observer 1/6/85).
10. In common usage *Ekhaya* means 'go home'. The context alluded to by the Prime Minister was that the unemployed youth, commonly blamed for the high incidence of urban crime, should return to their homes to cultivate the soil. Urban crime has flourished because many Mozambiquan refugees and *sidlanis* (youth gangs) live on the urban fringes.
11. The tractors were not donated and payments demanded from the chiefdoms were excessive. They were sent out well past ploughing time, and not all reached their destinations. Finally it was found that many never went beyond the transport authority yards where they were cannibalized for parts (Times of Swaziland, 18/12/85).

2 The Ntimane Irrigation Scheme

Introduction

This chapter provides a background account of the chiefdom of Ngwempisi and of the Ntimane Irrigation Scheme, outlining the routine operations of the scheme, and describing the manner in which particular values have come to be instrumental in influencing behaviour in an increasingly commercial environment.

The chiefdom of Ngwempisi

The Ngwempisi valley is a middle veldt area of approximately forty square kilometres, at an altitude of 750-1000 metres (Goudie, 1983, p.16). Ngwempisi is partially bounded by the Mtimane (a small stream), and the Ngwempisi Valley Road at the extreme eastern tip. The Ngwempisi river forms the southern boundary, and the Dwaleni-Mliba ridge road, the northern boundary. The western boundary, towards the Republic of South Africa, is roughly marked by the Mliba-Piet Retief road. A single road dissects the chiefdom. It intersects the main Dwaleni-Mliba road at Dwaleni and passes close to Ntimane, and the Chief's homestead at Mapokane, finally joining the Mliba road near Mliba.

The main physical characteristics of the Ngwempisi valley are shallow gray sand to sandy loam (lithosolic) on hard rock (hornblende biotite tonalite gneiss) (Government of Swaziland, 1979a, Murdoch, 1968). Smaller areas of deep red loams and reddish yellow sandy loams occur on higher river terraces (Government of Swaziland, 1976b, p. Annex 7, p.3).

Ngwempisi is characterized by rolling hills, broken by granite peaks and small rocky hills. The Ngwempisi River, one of the four major rivers of Swaziland, is fed by strong tributaries, including the Ngwempisana, the Towela, the Ntungulubi and the Mtimane. Overlooking the valley are two major peaks; to the south east looms Mount Hokwane, and to the east, Mount Dwaleni.

Ngwempisi is a good farming area with an average rainfall of 890-1270 mm, higher than average for middle veldt areas (Goudie, 1983, p.22). Summer is hot and wet, but the relative humidity is relieved by the cooling breezes from the surrounding high veldt. During the summer the countryside is green and lush and all the streams fill with water. Winter is cold and dry in Ngwempisi. The sparse pasture is brown and the ground becomes hard and dusty. During the day time the direct sun is pleasantly warm, but at night the temperatures drop dramatically, and the air becomes chill and frosty.

Homesteads dot the entire area, separated by maize fields which radiate concentrically from each. The homesteads are a mixture of traditional round mudhuts with thatch roofs, and the newer square cement block huts with corrugated iron roofs. In summer, which is the peak growing period, there is always much activity as people plough the soil, plant crops and hoe weeds. In winter, after the crops have been harvested, people retreat into the warmth of their kitchens, or sit in the sun. During the winter, the children are often on holidays and they can be seen playing in the maize fields, near the river, or about their homesteads.

The homesteads have a relaxed feel to them. The pace of life seems slow, and the demands of life appear minimal. The homesteads have no electricity and people go to sleep early and rise early. People walk to most places, always in an unhurried manner, and inevitably take time to stop and greet each other. Herdboys take the cattle out in the mornings and bring them home in the evenings so they can be milked. In the evenings, as the darkness descends, huts are lit by the weak yellow light of kerosene lamps. People move slowly and talk quietly.

In terms of material culture, Swazi are buying more and more western consumer items. Many homesteads now have lounge suites, beds, tables and chairs. Especially popular are aluminium pots and pans, so prized by many women. Also popular are china plates and steel cutlery, and plastic basins and buckets. Since the homesteads are agricultural centres, most homesteads will also have an array of hoes, spades, and shovels. The single furrow plough drawn by a span of oxen has always been in demand. Oxen are also used to pull sleds, which are used for carrying large quantities of wood, freshly harvested maize cobs, or water drums. The most obvious symbol of the degree to which the rural Swazi are attuned to western material culture, is the ever present plastic shopping bags, blowing in the wind, catching on the fences, and being eaten by goats.

The road running through Mapokane from Mliba to Dwaleni defines the centre of the chiefdom, and is the focus for most movement through the area. The road is used intermittently by a few government vehicles, occasional visitors, some local vehicles, and the twice weekly bus service. On the days when the bus is not passing, visitors leave the regular service at the Dwaleni turn off and walk the ten kilometres to Ngwempisi over steep hills. At all other times the road is the preserve of local people, as they walk to and from their work, or visit other homesteads. Children are always in evidence, especially in the mornings and afternoons, walking to and from the school at Mapokane. The only other traffic, apart from cattle herders and the occasional oxen drawn sled, is the tractor drawn maize mill which visits the area once a month.

Although only twenty kilometres from the Republic of South Africa, Ngwempisi was overlooked by Boer farmers who were more interested in high veldt grazing. Ngwempisi became a 'Native Area' after the land partition of 1907, and by 1932 was over stocked and over populated (Booth, 1983, p.23). As part of the Mliba District, Ngwempisi was targeted for rural development in 1956 by the colonial administration (Cruise and Holleman, 1964, p.50). With the advent of the Rural Development Areas Programme in 1970, the Mliba area was classified as an 'Intensive Rural Development Area' (Government of Swaziland, 1976b, p. Annex 7, p.1-14).

Ngwempisi is part of the Malamba Rural Development Area, and, although population statistics specifically for Ngwempisi are unavailable, the Malamba Rural Development Area includes approximately 2,054 homesteads and 21,950 people. Approximately eleven people are resident per homestead, of whom five are involved in 'farming' and two are classified 'absentee' (Government of Swaziland, 1985b, p.3-7).[1]

Although Ngwempisi constitutes approximately one-fifth of the Malamba Rural Development Area, about one-third of the population of the Malamba Rural Development Area lives in Ngwempisi. This population is concentrated on the western edge of the chiefdom of Ngwempisi, close to the main town of

Mliba (Goudie, 1983, p.42, 50). Mliba is a rapidly developing major regional centre, with a market, a small hospital, a hotel, a shop, a high school, a library, and a police station. The town also services the Mahlangatsha Rural Development Area, the High Veldt Ranch (part of a USAID beef improvement project), and other farms and projects.

Rural development in Ngwempisi

The implementation of rural development in Ngwempisi was the culmination of a variety of factors.[2] After the 1907 Land Partition Proclamation (see Chapter One), people in Ngwempisi were restricted in the land they could occupy while the population was still growing. Furthermore, because of the close proximity of Ngwempisi to South Africa, recruitment to the South African mines had always been a popular option. Further strain was placed on the available resources, when in the late 1930s the development of Usutu Pulp displaced many people who then came to live in Ngwempisi. The first development programme in Ngwempisi was the 1954 Native Lands Scheme, financed by the British Government, which defined specific areas for development (Booth, 1983, p.94), but had little effect in the area.

In 1968, a large Rural Development Area was established nearby (Davis, 1980, p.31; Funnell, 1982b, p.274-275), and then in 1978, half of that area, which included Ngwempisi, was re-classified as the Malamba Rural Development Area (Government of Swaziland, 1976b, p. Annex 7, p.1-14).

Currently, the Malamba Rural Development Area has a full complement of staff, and a substantial headquarters, with housing and vehicles for all personnel. Beyond satisfying the requirements of the organization itself, few other stated goals have been achieved (Government of Swaziland 1985, p.147). For example, no road construction has occurred, and road maintenance is sporadic and of poor quality. In Ngwempisi only one domestic water supply has been constructed, no government sponsored irrigation schemes have commenced, and there has been neither pasture improvement nor donga (gully) rehabilitation. Terracing and bush clearing have not been carried out and only one footbridge has been built over the Ngwempisi River.

When the Malamba Rural Development Area was incorporated, all pre-existing enterprises came under its umbrella. In Ngwempisi, these were the

Motane and Ntimane Irrigation Schemes, and the local farmers' association. The Malamba Rural Development Area encouraged the irrigation schemes to form committees, for liaison purposes, and gave them written constitutions. The irrigation farmers received few specific benefits. They obtained intermittent and often unreliable access to machinery, grants, credit and advice. Despite this, the farmers perceived other unintended benefits. Once Ntimane was established the farmers found they had to seek out other contacts. As a result, they have developed an awareness of the availability of credit, technical and agricultural assistance, and veterinary advice, outside the auspices of the Malamba Rural Development Area. An example of this is my reference (in Chapter One) to the rueful appreciation felt by the members towards Rural Development Area officials who, by denying the farmers access to the project trailer, taught the farmers the need for purchasing their own equipment. Another example concerned the large centralized maize storage facilities built at the Malamba Rural Development Area headquarters. Farmers were advised to grow maize and store it for sale in these bins. The system was never completed. The farmers quickly realized the importance of using their own small steel storage bins, and also chose to handle their own marketing, rather than rely on the officials to handle these transactions.

The Ntimane Irrigation Scheme

The largest and most commercially viable of the irrigation schemes is the Ntimane Irrigation Scheme. Ntimane [3] is four kilometres from the Chief's homestead along the valley road on the banks of the Mtimane river. Ntimane is a series of forty-five regularly shaped contoured gardens in well cultivated red loam, marked by sugar cane windbreaks and enclosed within a strong wire fence. The rows of cabbages, green peppers, carrots, tomatoes, eggplants and sweet and 'Irish' potatoes are neat and abundant, and in marked contrast to the hard pebbly ground of the surrounding countryside. The Mtimane cuts a deep track through the valley's edge and the gardens are concentrated near the river and extend away over a gradual slope towards the valley road.

From the valley road, a short spur road leads to the shed which is the focus of activity for the Ntimane farmers. All meetings are held there, cooperative resources are stored there, and it is the venue for all major social events associated with Ntimane. The shed is well situated, overlooking the gardens and allowing an excellent view over the gardens (see Figure 2.1).

Ntimane grew out of the initial efforts of a Mr Ndlandla, who in the 1940s, tired of working in the City Deep Mine ('Deepies') in Johannesburg, and decided to try growing vegetables at his homestead in Ngwempisi, where he tapped a very small stream for his irrigation needs. According to Mr Ndlandla, at first he carried his surplus to Mliba on his head, but found this to be too strenuous and, since his neighbours did not grow vegetables, he soon developed a local clientele by giving vegetables away. When demand increased, he began to sell his vegetables. This process he described to me, with some humour, as 'educating' his neighbours. Since that time he has satisfied all his financial needs from farming his gardens. When Ndlandla's commercial enterprise outstripped its water supply in the 1950s, he formed a partnership with a Mr Nxumalo, and they then both irrigated and farmed some of Nxumalo's land near the Mtimane, on the present site of Ntimane. The success of the operation was such that, by the early 1960s, the two men had formed an alliance with two brothers by the name of Msibi, and further developed their water supply and enlarged their gardens. About 1966, rapid expansion occurred when, with the support of the Chief, seventeen new members were enrolled in the scheme, and a two kilometre irrigation canal was built. In 1972, under the auspices of the Mahlangatsha Rural Development Area, funds became available to expand the cultivated area, and to construct a reservoir for water storage, which enabled the members to irrigate the larger area of the gardens. In 1978, with the advent of the Malamba Rural Development Area, a further section of land at Ntimane was developed and funds were supplied for the present farmers' shed to be built. At that point, the gardens reached their current size of forty-five gardens and thirty-five members.

Most recently, with some external financial assistance, the farmers have constructed a dam across the Mtimane and a three kilometre pipeline, and have doubled the size of the existing reservoir. From fees levied on members, they financed the purchase of a tractor, trailer and plough to be used by Ntimane members at the gardens. This machinery is also used as a source of revenue through hire for other purposes. So successful has this been that the farmers plan to buy another tractor specifically for hire.

Individual gardens range from 5000 square metres in the original sections, to 4500 square metres in the newer areas. The entire scheme covers twenty-one hectares in a hemispherical shape with a large gully running through the centre. The layout of the gardens is shown in Figure 2.1, and the major stages of development are shown in Figure 2.2.

Agricultural practices at the gardens are labour intensive and technologically unsophisticated. From the reservoir, water flows into the gardens from a large stopcock at the base of the wall, through small canals which feed the individual gardens. Large quantities of grass grow about the plot areas, and grass and weed abound in the canals. The taller 'thatch' grass is used to germinate seed, provide shade for seedlings, ripen frost damaged tomatoes, thatch homesteads, and line new canals. The weed in the older canals prevents erosion and obviates the need for concrete troughing. Communal nurseries for the propagation of seed are sited between the shed and the reservoir, although some members have private nurseries in their own gardens. Heavy ploughing is now done using a tractor, which has replaced oxen. Furrows are constructed manually, and seedlings are then transferred from the nurseries and planted two thirds of the way up the downhill side of the furrow, and individually fertilized with *kraal* (cattle pen) manure or chemical fertiliser.[4] Chemical pesticides and herbicides are regularly applied with privately owned knapsack spray packs, and farmers seem unconcerned by their probable exposure to toxins. Wire and sticks are used to stay tall plants, and harvesting is carried out by hand as produce is sold.

All marketing is the concern of the individual plot owners, and usually involves sales to local customers. External sales may involve 'middlemen' who transport the vegetables to markets in Swaziland or South Africa.

As members of Ntimane, the farmers are obliged to contribute to community work in addition to their own work. The farmers often state openly that this community work is vital to Ntimane's success. Once the need for community work has been agreed on, the chairman of the scheme formally allocates and supervises the work on the specified day. All members contribute to the community work according to the number of plots they use. This contribution, which takes the form of one worker per plot per session, is usually given in the four hours prior to lunch. Failure to contribute the necessary labour incurs an fine of one lilangeni per plot per work session. Fines are deposited in the cooperative bank account, and the record of work sessions serves as a basis for determining part of the entry fees of new members, calculated at the rate of one lilangeni per session.

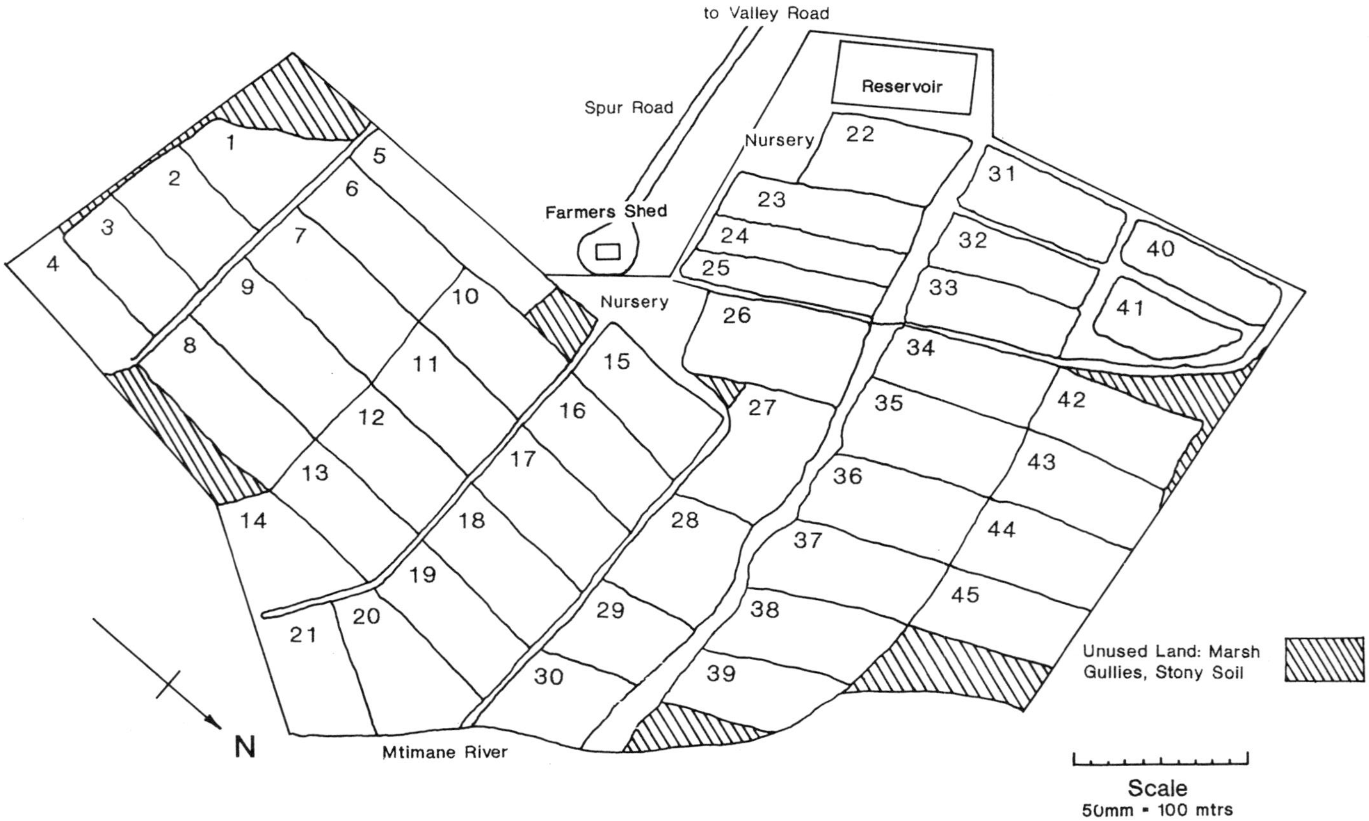

Figure 2.1 Ntimane Irrigation Scheme

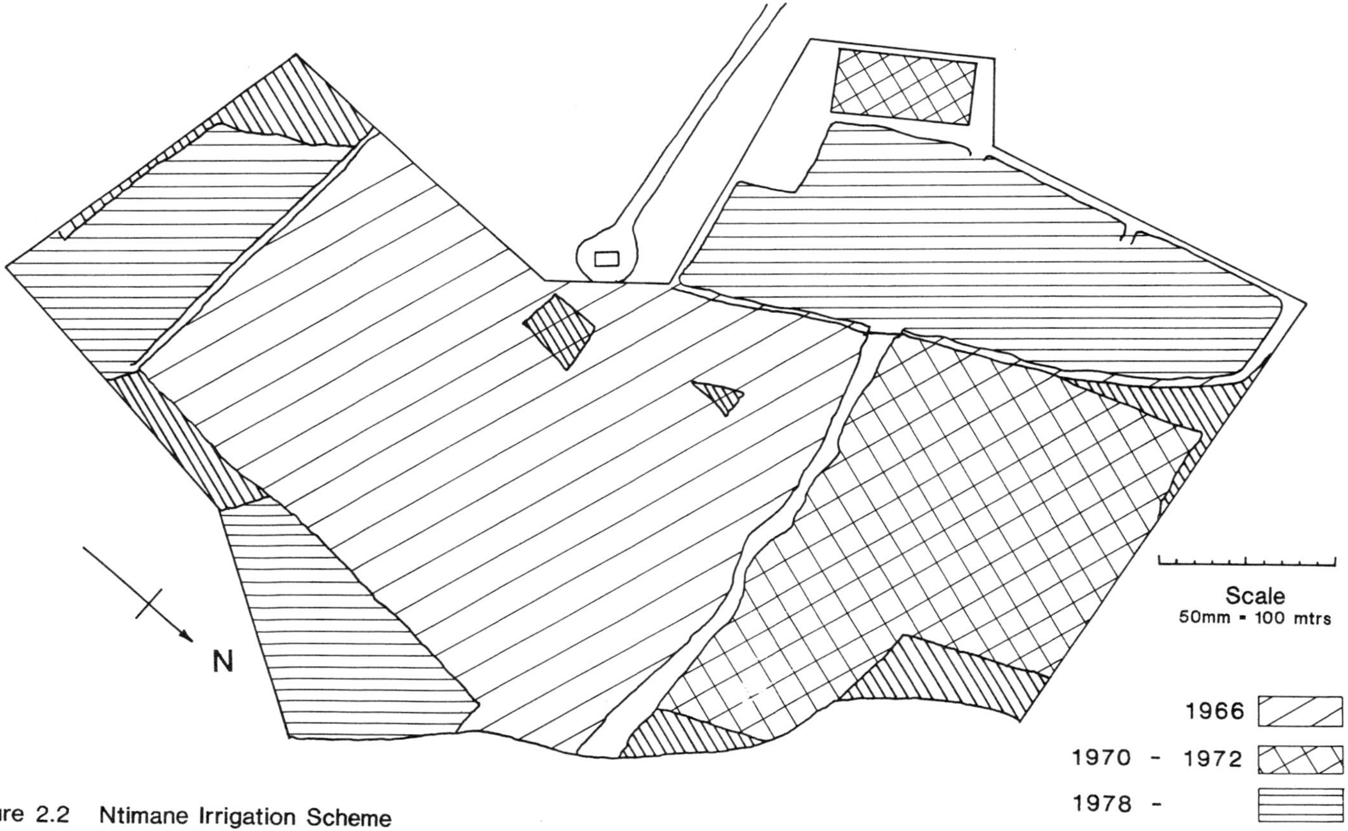

Figure 2.2 Ntimane Irrigation Scheme
Major Stages of Development 1966-1986

The plot owners and their families usually cultivate their own gardens with no obvious division of labour by gender. Some gardeners employ non-family labour. Daily workpractices are routine, and are coordinated with homestead duties. The farmers arrive at the gardens at about 9.30-10.00 am, and may spend a little time talking with one another. If community work is scheduled, they begin immediately and usually work until 1.30-2.00 pm. At this time food is eaten, farmers rest, announcements are made, and meetings may be held. From about 3.30 pm until well into the evening, the farmers work in their own gardens before returning to their homesteads.

Ntimane has an elected committee and a written constitution. The committee comprises the chairman, vice-chairman, secretary, treasurer and three ordinary members. The constitution was prepared by personnel from the Malamba Rural Development Area in 1978. Mr Ndlandla, the original farmer at Ntimane, became the foundation chairman, remaining in that position until 1982, when he retired in favour of a younger man. The new chairman, Mr Ntambo, is considered highly competent and is respected and trusted by all the members. He still works closely with his predecessor, but also relies heavily on support from the vice-chairman, Mr Mashinini. Formal administrative procedures are vague, and elections, which are decided by a show of hands, have generally favoured males for the top administrative positions. However, this is changing. Four of the seven committee members are women, and even outside Ntimane women now head both the pipeline and cattle dip committees.

Although only fifteen of the forty-five plots are registered solely in women's names, women maintain a proportionately higher profile at Ntimane than is reflected in plot ownership. The fifteen women who own plots usually work those plots alone, or with limited assistance from their children. Some other women assume primary responsibility for plots registered in their husbands' names. With few exceptions, the wives of most active plot owners work with their husbands in their gardens. Therefore the women involved at Ntimane are in a position to make major economic decisions for their families, and formulate major production and marketing strategies for their individual gardens in a manner inconsistent with traditional role models.

Becoming a member of Ntimane does not conform to the procedure outlined in the constitution. I was told by both the chairman and the former chairman that new members are chosen from a pool of potential candidates solely on the basis of their proven expertise as farmers, their ability to meet the capital outlay involved in joining Ntimane, and the extent to which they could contribute to the harmonious operations of the scheme. For example, Mr Mkholo currently heads the list of candidates for membership. He is a very experienced irrigation farmer, and has a successful farming record. Both the chairman and the former chairman favour him because he is successful and because they believe that he will work with the members of Ntimane in a harmonious atmosphere.

On the rare occasions when a plot becomes available, the new member must buy entry to the scheme. For example, if 1000 community work sessions have been carried out at one lilangeni per day, and a forty-fifth share of all equipment (tractor, trailer and plough) amounts to E200, a new member must pay an entry fee of E1200. Entry thus becomes progressively more expensive with time, but all members, on entry, have equal rights.

Political representation for the members of Ntimane is enhanced at both the local and the state levels by their relationship with the Chief, Prince Mkhosi Dlamini. At the local level, the Chief is nominally in charge of Ntimane, and theoretically adjudicates all disputes, though in practice he adjudicates only those the committee cannot resolve. The Chief has a personal stake in Ntimane, since two wives, his daughter-in-law and his brother are members. At the state level, the Chief has the power to approve local candidates for election to the *Tinkhundla* (Regional Assemblies), thereby ensuring that local members of parliament will probably be sympathetic to Ntimane.

Ntimane maintains an extensive network of contacts with both government and non-government organizations. The bulk of government contact is with officials representing the Ministry of Agriculture and Cooperatives, who oversee the activities of the Malamba Rural Development Area. Most contact is maintained by the Rural Development Area Extension Officer, who is personally well liked, and is invited to all social occasions.

Ntimane also maintains contacts with non-government organizations, the most significant of which are the U.S. Embassy in Mbabane, the Malkerns Agricultural Research Station, and the Unitarian Service Committee of Canada.[5] The farmers also patronise private firms which supply farm chemicals, seed, fertilizer, and milling facilities. These latter contacts they consider to be vital as they provide access to quality products from South Africa.

Ntimane is one of approximately fourteen cooperative schemes in Swaziland (De Vletter, 1986; Sithole and Testerink, 1988, p.4), and shares some common traits with these other schemes. Unfortunately, little information concerning these other schemes is available on which to base comparisons. De Vletter (1986) conducted a survey of the cfourteen schemes on Swazi Nation Land over the period of December 1985 to March 1986, and he reached the following conclusions:

1 The majority of schemes are cooperatives with individual farmers farming individual gardens.

2 Though not necessarily directly involved, the local chiefs are instrumental in the operation of successful schemes.

3 All the schemes have elected committees consisting of a chairman, vice-chairman, secretary, treasurer and three ordinary members.

4 Most schemes have disproportionately high numbers of women involved in the committee, when compared to actual membership within the scheme.

5 Most non-government initiated schemes restrict membership to selected, well financed applicants.

6 Seasonal labour is hired.

7 Work parties are compulsory, with fines for non-attendance.

8 The main cause of eviction is the non-cultivation of a plot by a member.

Ntimane largely conforms to De Vletter's description. The dominant emphasis among members at Ntimane is on individual responsibility and initiative, with strong leadership, a definite work ethic, and a determined commitment to commercial farming. Ntimane is somewhat larger than the twenty-three member average suggested by Sithole and Testerink (1988, p.4). A variety of factors provided indications to me that Ntimane could be regarded as one of the most successful schemes of its kind in Swaziland. One factor concerns the way in which outside agencies use Ntimane as a show case scheme. For example, during my stay from August 1985 to August 1986, two major visits occurred. The first was made by the USAID chief for Africa, who was on his first visit to Swaziland, and the second was made by the representatives of CYMMET (The International Maize Research Organisation). Other informal visits were frequently made by personnel from the Malkerns Research Station. The Unitarian Service Committee of Canada has made specific reference in the media to Ntimane as a well managed and successful scheme (Swazi Observer 23/4/86), and numerous government and non-government personnel made references to the same effect to me.

The Real Farmers at Ntimane

The majority of Real Farmers identified by me in Ngwempisi are involved at Ntimane. These Real Farmers have certain characteristics in common (see Chapter One). In particular, they emphasize the resolution of issues through consensus, and they espouse a work ethic, which they describe in terms of christian values. In the operation of Ntimane, these characteristics are evident.

Consensus Decision making at Ntimane operates on two levels. The first, the 'democratic' level, involves issues being put to the committee and ratified at a general meeting which, although dominated by prominent Real Farmers, runs smoothly and without signs of dissent. This form of decision making only ever occurs when the issues are logistical and/or administrative, and have no potential for division. Meetings I attended, which were run in this fashion, dealt with routine matters such as scheduling water supplies, organizing social occasions, and deciding on necessary community work. The most noticeable feature of this level of decision making is the conservative demeanour of the members, who act in a deferential and respectful manner to their leaders. This behaviour is in accordance with formal behaviour in any public forum in Swaziland. In this situation, leaders are able to exercise those qualities so admired by the Swazi; eloquence, verbal courage, and charisma.

An example of an individual with all these characteristics is the former chairman, Mr Ndlandla, who uses a combination of humour and incisive criticism to convince the members to support him. Mr Ndlandla's ability to sway the opinion of the members rests not only on his reasoned arguments but on his force of character, an attribute of leaders which is widely admired in Swaziland. King Sobhuza II was generally recognized as having this attribute. Kuper (1978) described Sobhuza II in the following terms:

> I observed a man of medium height and build, with a strong face, deep brown in complexion, a wide forehead, broad flared nose, high cheek bones, bright, deep set penetrating eyes, a short, natural, untrimmed beard, and smiling mouth. But it was not the build or features that set him apart as much as a general impression of animated intensity and personal dignity (Kuper, 1978, p.1).

The most prominent of the Real Farmers, like the loved and admired King Sobhuza II, are such people. According to popular belief, the strength of such people can be seen in their eyes. People told me that it is obvious in the eyes of people, when they are people of strength. While Mr Ndlandla embodies these characteristics, female Real Farmers, though constrained by their traditional reluctance to speak publicly, are noted for their animated personalities and determined natures. Humour is greatly appreciated in such people and, seems to be used to defuse potentially divisive situations. Such qualities in Real Farmers are shared by Real People generally. The recognition of such qualities in important people reached its height with the coronation of King Mswati III, when people referred to him as having 'his father's eyes', and gave him the sobriquet 'Fire Eyes'.

The second form of decision making at Ntimane is on a more paternalistic level, in which the Real Farmers take advantage of the respect and deference they are accorded by other members, to restrict any chance of dissent in public debate. Meetings held under these circumstances are used to confirm certain pre-determined goals, and achieving consensus rests directly on prior intensive political activity. In this instance, issues arise privately and become part of informal general conversation during work sessions and regular siestas. As these issues take form, the prominent Real Farmers articulate and refine opinions, which are then often taken up by the committee for adjudication and promulgation. Most of the decisions are made in this manner. For example, the decision to buy the trailer for the tractor occurred only after the farmers had tried to hire a trailer from the Malamba Rural Development Area. Though they were successful on some occasions, the trailer often never arrived, and the farmers would wait for half a day to start community work only to return home and re-convene the next day. General frustration at this situation boiled over, and the prominent Real Farmers directed that frustration into an active fund raising programme by the committee to buy a trailer. They did so by waiting until the members were discussing or complaining about the inefficiency of the Malamba Rural Development Area personnel, and then suggested that perhaps the committee might consider the problem. The chairman then replied that he would take it to the committee and the process was instigated. Since, in that case, all members could easily see a need for their own trailer, there was little opposition to the idea, and the process was handled expediently.

The most intense political activity surrounds potentially divisive issues. As such an issue gains momentum in informal general discussion, prominent Real Farmers will introduce it into informal siesta discussions in a general, non-threatening manner, at a time when the person who is the subject of the issue is not present. A consensus of opinion is then usually quickly reached.

Having established consensus, the first action of the Real Farmers is to try to get kin to influence the offender to modify his or her behaviour. Usually this is all that is required, since strong kin relationships provide a variety of possible sanctions. Minor infringements of cooperative rules may also be dealt with by a prominent Real Farmer directly rebuking someone, in a good humoured manner, during the siesta.

For example, on one occasion a headman had made a contract on behalf of Ntimane which he found he could not fulfil. A kinswoman took up that contract and saved the scheme some embarrassment. Subsequently, she was privately rebuked for her actions by her absentee urban salary earning husband. On the next public occasion, a prominent Real Farmer praised the woman for her initiative, suggested that others could emulate her, and thanked her for saving the cooperative embarrassment, despite the fact that they were all aware of her husband's annoyance. Since on this occasion her husband was present, this constituted a very public rebuke for him, and his shame was exacerbated by the appreciative applause offered to his wife. It also constituted a significant diversion from traditional notions of female respect for the authority of men. In particular, this incident illustrated a definite move away from the circumstance where a wife's concurrence with the wishes of her husband was considered sacrosanct (Kuper, 1963, p.26).

Serious infringements involve more intense procedures. One man failed to cultivate his garden and, according to the constitution, could have been expelled. Rather than officially expel him and risk acrimony, the Real

Farmers took advantage of a witchcraft accusation directed against the man from a member of the general community. They then encouraged the man's kinsman to ask the offender for tenure of the plot, assuring that kinsman that he had committee support. Faced with sanctions from the Chief over the witchcraft accusation, the man responded well to the 'spontaneous' offer from his kinsman to manage his garden and he left Ntimane. After an appropriate period the Ntimane members recognized the right of the new tenant to full ownership of the plot. At no point did the issue become an open topic subject to formal or public discussion. Significantly, it was the witchcraft accusation which provided the Real Farmers with the moral rectitude to justify the resulting machinations. This is interesting because two other members of Ntimane are recognized as *batsakatsi* (wizards), but have not lost their rights to own plots. Clearly, Real Farmers are using witchcraft accusations only when it suits them, to achieve a particular result.

This process of sorting out issues in a non-public way is exemplified by the methods by which prominent Real Farmers allow certain potential members to gain a toe hold in the scheme. Formal induction usually involves a waiting list, a transfer of fees, and capital investment in plant. Instead of being placed on the list, some enterprising individuals are allowed to make arrangements with individual plot owners to cultivate a portion of their plots. Owing to pressure on plot owners to use their entire plot areas, some members begin to have difficulty cultivating their entire plots as they get older. Prominent Real Farmers will encourage those members to allow selected non-members to cultivate part of their plots, and so establish some precedent of involvement for the future.

This procedure is clear in the particular case of plots fifteen and sixteen (see Figure 2.1). These plots are used by a husband and wife, both members, who are getting too old to produce commercially, and prefer to use their gardens for subsistence. They have allowed the sister of the prominent Real Farmer on plot four to use part of plot sixteen (the wife's plot). This woman is a hard working widow, without a homestead, who has come to live with her brother in Ngwempisi. Plot fifteen is shared by the man with his daughter from his second wife who is a non-member. His daughter will probably inherit her father's garden. The hard working widow, having established herself by successfully working the old woman's plot, will be ready for formal entry when another garden becomes available, or when Ntimane expands, or perhaps even when the old woman dies, although there may be claims from other relatives of the old woman.

In both cases these issues never reached the committee level, being dealt with in general conversation, and with decisions being reached through consensus. There are emotive issues implicit in this process of preferential entry, as many people want to join Ntimane. When the issues are resolved they tend to rest on prominent Real Farmers, speaking on behalf of the plot owners and representatives of the applicant, putting forward arguments for allowing the arrangement to occur, and gaining general assent. If an unpopular application is made, it simply never gets discussed and, having no support, will not be successful.

Within the social organization of the Ntimane Irrigation Scheme, formal democratic processes as envisaged by the Rural Development Areas Programme are limited to those issues that are mechanistic and devoid of political nuance, such as deciding on work rosters. The most important issues are decided in an on going and intense dialogue among members at various levels of formality. Superimposed on this process is the influence of the prominent Real Farmers, who moderate and direct the exposition of issues in a focused manner, which ensures consensus and harmony for the entire cooperative and avoids the pitfalls of faction politics, while strongly constraining the power of the elected committee.

The members of Ntimane are not perturbed by the way the Real Farmers limit democratic processes. Some younger people expressed to me a degree of frustration over their relative lack of power, but did so while accepting the principle that the Real Farmers have an inherent right to wield such influence. This is a principle that has widespread acceptance, even at the national level, and is most evident in the way King Sobhuza II was seen, and in the manner in which people currently regard King Mswati III. King Sobhuza II actively promoted paternalism as an appropriate cultural reality for the Swazi. This attitude to authority is still current in the pride and comfort people find in a paternalistic and caring authority, and it provides a strong role model for the Real Farmers to adopt in public discourse.

The acceptance of paternal political authority by the general public was clearly demonstrated between 1968-1973, when the Swazi experiment with a parliamentary democracy was perceived by many people to be ugly and divisive. Specifically, the Westminster system of parliamentary democracy was considered divisive because of the manner in which factionalism within, and between parties, was marked by open and aggressive rhetoric between principals. Informants who recounted their experiences of the period to me were convinced that the political process was linked to witchcraft, and registered real distaste when referring to the manner in which the supporters of various parties had argued with one another. Their distaste was based on a sense of certainty that witchcraft would inevitably follow such behaviour.

As the example of the experiment in democracy demonstrates, the Real Farmers at Ntimane are able to act with the assurance of knowing that people will avoid public divisiveness if possible. Clearly, though, witchcraft plays a part in ensuring that leaders act fairly towards their followers, and prevents too rampant an abuse of power by leaders. Consensus politics, thus achieved with internal checks and balances, represents the most effective form of political action for the Real Farmers of Ntimane.

The influence wielded by the Real Farmers rests on their ability to accept and resolve all the ambiguities that arise from the circumstances they, and the other members of Ntimane, find themselves in. They are able to reassure the other members of Ntimane, and the members of the general community, that they do not threaten the common good through their actions. In this process, they are able to constantly adjust their own perceptions of tradition to accommodate the demands of modernity, and at the same time convince others that their actions are appropriate. This is well illustrated by reference to the neighbouring irrigation schemes of Motane and Mzimneni. Motane is essentially a smaller version of Ntimane, having been stimulated by the success of Ntimane, which was widely recognized by the people of the chiefdom. The dynamics of Motane mirror those at Ntimane, and political processes are dominated by a prominent Real Farmer who expressed to me opinions and attitudes similar to those of the Real Farmers at Ntimane.

The Mzimneni Resettlement Scheme provides a noticeable contrast with Ntimane in terms of the manner in which the members of the scheme manage their affairs. The example of Mzimneni also provides a clear comparison with Ntimane in a way that provides insights into why Ntimane functions as well as it does. Mzimneni commenced in 1978, under the auspices of the Rural Development Areas Programme, and USAID. People were resettled to make way for the High Veldt Ranch, a cattle breeding and fattening complex. An American adviser was involved in the planning of the new irrigation scheme over a three year period. He controlled the use of funds and the design of the scheme. The total area used amounts to 674 acres, [6] and is the site of a former private farm. Water for irrigation and for domestic use is drawn from the Ngwempisi River through a twelve inch pipeline. The twelve inch pipe empties into a canal which fills the dam for irrigation. Twenty-four hours of pumping provides forty-eight hours of irrigation. Every homestead has a water tap on a polythene pipe for domestic needs. The field areas were levelled mechanically by tractor drawn machines, and electricity was brought in to run the pump.

Each homestead was allocated a specific area and given two acres of irrigated garden, and nine acres of maize fields, as well as access to the grazing area, which is fenced and adjoins the High Veldt Ranch lands. Thirty-four homesteads were relocated to Mzimneni on a voluntary basis. The aim of the scheme was to grow vegetables for commercial production, and to grow maize for private consumption. There was a E10 joining fee, and, for an initial period, electricity was free. All infrastructure was supplied free of charge by USAID and the Rural Development Areas Programme.

By 1986/87, the scheme was clearly unsuccessful since none of its goals had been achieved. Members to whom I spoke gave three major reasons for this; there is no water for irrigation, they are too remote from the Mliba market, and they have no land for their sons. Certainly, they encountered infrastructural problems almost immediately after the scheme was commenced, but none of these problems was insurmountable. The electric pump, which serviced the dam, was destroyed in the cyclone of 1983 and was taken away, never to be replaced. Eventually the Swazi Electricity Board removed the transformer from the end of the line, which compounded their problems still further. Access to Mliba is difficult. Although the scheme is only two kilometres from Mliba in a direct line, the road journey is about thirty kilometres. However the Ngwempisi River at the foot of the scheme is very low, and a vehicle ford could easily be built. The shortage of land for sons is also real, but no more pronounced than elsewhere.

The farmers blame their lack of success on recent occurrences, such as storm damage sustained in Cyclone Domoina in 1983, and also on the loss of the pump, and the hidden costs associated with electricity. Electricity charges have been backdated to the beginning of the scheme and the farmers feel that they have been mislead, because they didn't know they would have to pay for electricity. Complaints to the Swazi Government resulted in the government paying half of the large bill, and then imposing a monthly fee. This fee was initially E150 but, by 1986, had become E200 per month for the entire scheme. The farmers told me that these charges were crippling them. The farmers also said that they were misled by the planners of the scheme, who informed them that it was King Sobhuza II's wish that they leave their homes and move to Mzimneni. They place great emphasis on this encouragement, and speak of the occasion in such a manner as to suggest that they had genuinely believed that King Sobhuza II was actually aware of their situation, and not that the king's wishes had been used as a ploy to get them to move.

I concluded that the extremity of their situation is not due to the recent breakdown in infrastructure. These disasters merely compounded an already bad situation. None of these problems would deter the farmers of either Ntimane or Motane. It seems that the problems at Mzimneni became evident soon after the scheme began, when the farmers failed to begin producing vegetables. This had been the situation for four to five years before the pump was destroyed.

The impression I gained was that the farmers use their infrastructural misfortunes to explain their lack of success. The committee members seem to lack motivation and there are no prominent individuals whom the people can identify as Real Farmers. Issues remain unresolved for long periods of time. For example, they have not resolved such basic issues as how homestead and garden inheritance is to be handled. They do not have a constitution, and simply say that they had not been given one eight years before when they commenced operations. This lack of a constitution certainly makes any resolution of issues difficult. Motivated members are effectively being short circuited by the unmotivated members, as they do not have grounds on which to control membership, or to enforce commitment to either cultivation, or to the general goals of the scheme.

The difficulties in getting members to commit themselves are, I concluded, compounded by a lack of selectivity in admitting members to the scheme. This situation originally occurred because, since the population was relocated, all the people are members regardless of whether or not they wish to farm commercially. None of the members had any previous experience of one another as commercial farmers. Gardens are allocated to homesteads, and are controlled by the males, although women supply much of the labour. There are no women members and, although women are represented at the committee level, it was made clear to me that women cannot be plot owners. This is in complete contradiction to the dynamics of the successful schemes, where so much emphasis is placed on rewarding female labour with land and political rights. This latter point is a crucial point of difference between the successful Ntimane and Motane schemes, and Mzimneni. When I broached the subject of women members, the men at Mzimneni argued that such rights would impugn tradition, reflecting a distinct inflexibility on the part of these men when compared with men at Ntimane.

The Mzimneni members are also handicapped by their lack of linkages with traditional authorities. When they were given autonomy from their former chief in Mahlangatsha, they were thrown into a position of reliance on the officials of the Mahlangatsha Rural Development Area. When the officials failed to deliver on their promises, the Mzimneni members sought help from their former chief. That Chief responded by trying to settle more people on Mzimneni land. The members told me that this chief is not progressive enough, and that they cannot rely on the officials of the Mahlangatsha Rural Development Area, so they are seeking incorporation into the Ngwempisi chiefdom. I concluded that, while they are impressed with the record of the two irrigation schemes in Ngwempisi, they do not understand the need to change in their own basic attitudes to commercial production and the role of women in that process. There was also, in the comments of Ngwempisi Real Farmers, a reluctance to become involved with Mzimneni, which could be attributed to a basic clash of values between the entrenched and inflexible attitudes of the Mzimneni men, and the versatility of the Real Farmers.

Further differences are apparent between the various schemes, in that the Mzimneni members express no interest in being involved with high profile bodies such as the *Imisumphe* (Land Management Committees) since, as they argue, all the land has been allocated anyway, and they have nothing to gain. The members at Mzimneni are alienated, without the will to use their autonomy, without either a commercial or political profile, and without the Real Farmers who could delineate the problems and plan resolutions. Mzimneni foundered on the ambiguities and uncertainties of democracy and opportunity over a short time. In particular, the members of Mzimneni have no process by which decisions are made to achieve corporate goals. Currently, they expect leadership from either a chief or a bureaucrat. They do not realize that they must have leadership from within their own ranks to achieve success. As the members of Ntimane have demonstrated, the involvement of the chiefs or bureaucrats is best exercised through the Real Farmers.

Christianity and work ethics The basis of consensus among the members at Ntimane rests on the common acceptance of a work ethic. This work ethic falls into a range of behaviours commonly referred to by the Swazi in their everyday life, and in the media, as 'christian' behaviour. I believe that this range of behaviours constitutes a work ethic similar to that, which in western tradition is commonly referred to as the 'Protestant work ethic'. Furthermore, many similarities between western and Swazi ethics are evident in the relationship between commercial success and christian values. Tawney (1938) describes this relationship in the following terms: 'By a fortunate dispensation, the virtues enjoined on christians - diligence, moderation, sobriety, thrift - are the very qualities most conducive to commercial success' (Tawney, 1938, p.244).

The members of the Ntimane Irrigation scheme are predominantly members of the Zionist Churches of South Africa. This is a powerful, though fragmented, fundamentalist religion with a long history in Southern Africa, which became dominant in Swaziland from the 1930s onward (Kuper, 1963, p.67; Marwick, 1966, p.249). The Zionist Churches are a truly syncretic element in Swazi culture, and one which allows an easy reconciliation of tradition with modernity. Kuper describes the advent of the Church in these terms:

> ... the proselytizing influence of all white missions virtually came to a standstill in the late 1930's, when the nativistic African Separatist movement boomed on an upsurge of nationalism. Separatist 'Zionist' leaders consulted with Sobhuza and decided to form the Swazi National Church, with a flexible dogma and a great tolerance of custom. Sobhuza was thus to be ritually entrenched, both as head of the traditional ancestral cult and as priest king of a new faith (Kuper, 1963, p.68).

By 1963, forty-seven per cent of the total population claimed to be christian (Hughes, 1964a, p.151), and the involvement of zionist churches in traditional national ceremonies had given a legitimacy to zionism that had not previously existed in Swaziland. Sundkler (1976) states:

> If, as we have suggested, a spiritual vacuum developed in the 1930's, this was filled by the Zionists, particularly through the imaginative decision of Bishop Stephen Nkonyane to attend, in his episcopal garb, the national *Incwala* festival and similar great occasions in the life of the nation (1976, p.241).

The zionists emphasized unity with Swazi cultural practice, easily incorporating polygyny, the ancestor cult and the *Nkulunkulu* (the Great Great Being). Zionism occurred at the same time as pressures on land tenure became acute. Furthermore, zionism was an indigenous, highly flexible belief system, and an alternative to the existing European dominated churches.

The growth of a work ethic among the Swazi is expressed in public discourse as both 'christian' and 'traditional' with no sense of ambiguity, for christian and Swazi identities are not considered to be exclusive. The role of expatriates in Swaziland throws christian beliefs into an interesting light for, in expressing the need for 'christian' values, the Real Farmers will often refer to expatriate role models who are particularly admired for their business management. For example, one headman referred to the common administrative problems with Swazi bureaucracy as the 'Swazi disease', and referred to bureaucrats as being:

> ... incapable of getting the job done. They don't seem able to push [to achieve a result]. They do not care that we need something desperately, all we can do is beg. If we visit them they say 'we will look into it', but then nothing happens. I believe eight out of ten Swazi suffer from this disease [the inability to complete the job] and I believe Europeans are much better at this. If Portch was still here the pipeline would be finished.[7]

In general dialogue the relationship between 'christian' and 'European' is demonstrated in the term *baya-european* (lit., 'to be like a European'), and is often expressed as a compliment. To have a 'christian wedding' is to be married in the European style, whether or not a church service is involved. Because of the close correlation between 'christian' and 'European', many values espoused by Europeans concerning work and work ethics are associated with 'christian' ethics in the minds of Swazi.

This is an artifact of a peculiar situation in which Swazi only have contact with certain types of Europeans. These Europeans are generally prosperous South Africans or educated, middle class expatriates. Understandably, the Swazi have the impression that all Europeans are like this select group, and they admire them. For example, one man explained to me how he came to have the name '*Mngisi*'(lit., 'The Englishman'):

> My mother worked for an Englishman and she greatly admired him, for he was a kind and hardworking man. She named me Mngisi hoping that I would become like him.

The Real Farmers closely relate the economic success of Europeans to their ability to organize labour, and to their 'christian' values. These values are thrift, work, and sobriety. Real Farmers often express disparagement for their fellow Swazi who drink to excess, and they admire those who work hard and manage resources and finances in an acceptable manner. Christianity is always a part of everyday affairs at Ntimane. One of the original members of the scheme is also a zionist priest and is linked to the royal family though the marriage of his sister. He leads the prayers that preface all community meetings and, by his presence among the prominent Real Farmers, he reinforces the moral and religious legitimacy of their actions in the minds of other people.

Despite the emphasis on christianity among Real Farmers, non-christians who are members of Ntimane, do not suffer discrimination. In many ways, the zionist Real Farmers resist christian fundamentalism that excludes Swazi belief. For example, they resist the activities of the Watchtower Sect (Jehovah's Witnesses), who, according to local zionists, try to drive the Swazi to a literal biblical interpretation, and provoke strife in the chiefdom. The members of the Watchtower Sect agitate by attacking the authority of the Chief, and by arguing that belief in the *emadloti* (ancestors) is a form of satanism. This latter attack is particularly significant, for zionism recognizes the role of the ancestors in providing role models for everyday life.

For the Real Farmers, the great strength of zionism lies in the zionist emphasis on authority and respect. Furthermore, zionism is now a part of the national ceremonial life of the kingdom. This enhances the legitimacy of zionist principles among the general public, and provides a basic reference point by which the Real Farmers may encourage unity. Zionist principles, as expressed by Real Farmers, are further enhanced by the degree to which prominent non-christians, such as the Chief, resort to zionist priests for help with certain community ceremonials. For example, in November 1985, when the rains had not yet arrived, the Chief gathered his subjects on the hill behind his homestead, where they all prayed for rain in a zionist ceremony.

Conclusion

Ntimane and Ngwempisi are intricately related. While it is evident from the constitution of Ntimane that the scheme enjoys control over tenure, this is limited by the continuing influence of the chief. Quite clearly, Ntimane has become the focus of activity in that sector of Ngwempisi, and many initiatives affecting the entire chiefdom originate among Ntimane members. It is there that particular values are articulated, and are instrumental in encouraging social change in the general community.

The opportunities offered by the Rural Development Areas Programme have enabled the Real Farmers to relate christian work ethics to commercial success, and ultimately, to the ownership of commodities. In the following chapters this relationship will be demonstrated in the way the Real Farmers influence the rights other people in the community have to gain access to the common resources of the chiefdom.

Notes

1. It should be noted that the definitions of 'homestead', 'absentee' and 'farming' are vague and unclear. De Vletter's rural homestead survey from 1983 indicates a mean homestead population for high veldt regions of 10.2 and a Rural Development Area figure of 9.5. This indicates that the ministry figures are in general agreement with De Vletter's as to the range of average homestead population (De Vletter, 1983, p.18).
2. 'It was during the succeeding period of Transvaal administration (1895-1899) that the Swazi first experienced the impact of economic deprivation. Part was caused by nature, and part by man. Rinderpest struck and decimated the cattle herds, beginning in 1894. In 1898

began the collection of tax, with the dual aim of extracting surplus to fund Boer administration and of providing induced wage labour (the tax was payable in cash) for local capital and for the gold mines on the Witwatersrand in the Transvaal' (Booth, 1983, p.16).

3. The Ntimane Irrigation Scheme is routinely referred to as 'Ntimane'. Throughout this book, whenever I use this designation, it refers to the Ntimane Irrigation Scheme.
4. As the water level drops it lies against the downhill side of the gully which is the last point to dry out. This technique ensures the roots are at the optimum level to draw water.
5. The farmers were given E14,500 (10,000 USD) from the U.S. Embassy, E32,000 from the Swazi Government and supplied E6440 themselves towards the cost of the pipeline they built. The USCC loaned E22,000 at low interest to buy the Massey Ferguson 260 tractor. Malkerns is involved with soil testing as the soil is acidic and needs liming.
6. This is broken down into sixty-eight acres of garden, 306 acres of maize field and 300 acres of grazing land.
7. Nick Portch was the Scottish engineer in charge of the Irrigation Department, and was the man responsible for the design of the pipeline.

3 Land tenure

Introduction

This chapter examines the question of land tenure in Ngwempisi, and how the Real Farmers are reacting to looming land scarcity. This issue has two parts. First, the Real Farmers are exercising significant influence over the procedures by which members of the general community acquire and use land. Second, previous rights of usufruct are changing to accommodate the emergence of private landed property and, in this process, giving allegiance in return for land is de-emphasized in favour of filiation in conjugal families.

Both of these issues are linked through the involvement of the Real Farmers. They have acquired proprietorial rights over land at Ntimane, through the Rural Development Areas Programme, and have contrasted these rights with those of usufruct, which they simultaneously hold at their homesteads. While the farmers of Ntimane have intensified production at Ntimane, they have deliberately failed to do so on their homesteads. This chapter will show that the Real Farmers are reacting to the threat of scarce land by trying to gain proprietorial rights at their homesteads.

Land tenure in Ngwempisi

In Chapter One, it was suggested that the house property complex was an appropriate model for envisaging traditional land tenure rights. Certainly, in Ngwempisi, rights to land in such terms have always been expressed clearly and unambiguously as a matter in which birthright is the main determinant for eligibility to give allegiance, and thus gain land. However, the system of land tenure, and land use, in Ngwempisi is threatened by a nation-wide pattern of growing population levels [1] and severe erosion.[2]

At the national level, concerns over the scarcity of land have resulted in inter-governmental negotiations with the Republic of South Africa over the KaNgwane disputed areas,[3] while within Swaziland there have been increasing numbers of disputes both within, and between, chiefdoms over pieces of fertile land (Kuper, 1978, p.111; Times of Swaziland, 28/10/87). These disputes are part of a fundamental change in attitude to land on the part of people who, in the distant past, had taken the availability of land for granted. This awareness on the part of rural dwellers has grown into a two-sided equation in which the interests of chiefs, in squeezing more retainers into their chiefdoms, must be balanced against the desire of their retainers to have enough land for their own use. This conflict of interests between the chiefs and their retainers confronts a long standing system by which chiefs enjoyed an enhanced status if they had many retainers.

This inherent conflict of interests is especially noticeable in the densely populated high veldt areas of Swaziland, where chiefs welcome agricultural intensification schemes as a means of accommodating growing numbers of retainers. This trend has encouraged some chiefs to become very active in promoting the needs of their chiefdoms. For example, Mnikwa Dlamini is the chief of the heavily populated northern Hhohho area. His attempts to seek assistance are constantly reported in the media, and he is often in trouble for trying to assert *kumemetela* (chiefly authority) over areas outside his jurisdiction. The chief has even defended the illegal cultivation of *dagga* (marijuana) by his subjects whom he claims are justified in growing this profitable crop (Times of Swaziland, 24/7/85, 1/8/85, 24/8/85, 7/10/85, 9/10/85; 31/10/85).

Such is the degree of competition engendered in competing for land that the process has been, and continues to be, marked by incidences of *muti* murder (ritual murder)[4] (Miller, 1933; Times of Swaziland, 1/10/85, 26/11/85, 27/11/85, 7/1/86, 1/11/86). Such are the pressures in areas such as Hhohho that the practice of ritual murder has not been limited to commoners (Times of Swaziland, 7/1/86; 30/4/86). In other parts of Southern Africa similar pressures have evoked similar responses (Mair, 1969; Marwick, 1965).

In Ngwempisi, competition for land is likewise intense, though not to the degree of the more heavily populated areas of Swaziland. The Chief, Prince Mkhosi Dlamini, retains control over the allocation of land and acts with the advice of his Land Management Committees, made up of leading commoners who have a good knowledge of the area in which they live. Like all chiefs, the chief of Ngwempisi has an interest in acquiring more retainers, and promotes agricultural intensification as a means of achieving this goal. He has stated:

> It is a big problem [land scarcity] ... children are growing up who want to give allegiance. It is a big problem ... what is happening is that everyone wants to have a small homestead instead of big ones ... but they must be able to get the land. The first priority and the first thing is to take from the foreigners, the soil is our greatest asset. Now we should take the methods of farming from the foreigners and we should use them. For everyway we go left or right we take our food from the soil. So our soil is our greatest asset. For example the Chinese [Taiwan delegation in Swaziland] are working day and night. This shows that even in their own country where they come from they know how to work and they work like this.

The Chief was determined to encourage agricultural production as a means of accommodating the growing local population. He was supported in this by his commonality of purpose with some of the policy goals of the Rural Development Areas Programme on the one hand, and the ambitions of incipient commercial farming interests on the other. However, an essential point of division soon emerged, and the Chief came to occupy a middle ground between the formal Rural Development Areas Programme policy, and the complex interests and stratagems of the Real Farmers. For example, the Ngwempisi Rural Development Area personnel wanted to separate grazing areas from living areas throughout the chiefdom. The Chief supported this strategy in principle but, since it involved widespread resettlements, he agreed only to a limited programme. Further opposition came from a variety of people who were also disturbed by the knowledge that resettlements also meant higher density settlement on the non-grazing areas.

The Chief's role becomes even more complex as he seeks to reconcile his personal ambition for more retainers with the evidence of growing population levels, the wishes of the central government, as expressed through the Rural Development Areas Programme, and localized tensions implicit in competition for land. The Chief needs to optimize the relationship between population and the productive capacity of the available land, if he is to avoid discontent. By endorsing intensive irrigation schemes, such as Ntimane, which occupy so little of the land in Ngwempisi, the Chief reconciles his need to have land available for those who give allegiance, with his retainers' needs for sufficient land.

The Real Farmers recognize that both objectives involve basic changes in attitude among the people of the chiefdom. Encouraging these changes of attitude, as they do, has particular implications for the way the people consider their rights to land, and their attitudes to labour. These changes are also particularly noticeable in the way they affect the status of women.

In fact, evidence of change beyond Ngwempisi has already been noted, particularly in the areas of women's labour and rights to land (Armstrong, 1985, p.18). Armstrong attributes the acquisition of female rights to land to male disinterest in commercial farming on Swazi Nation Land. To the contrary, in Ngwempisi, it is the Real Farmers who, recognizing the value of female labour and organizational skills, have encouraged women's access to land. Commercial farming is labour intensive but many homesteads have only limited labour available. In response, the Real Farmers are adopting 'women's work' (hoeing, planting, weeding and cropping). By doing this they are giving such work a heightened moral value, which has further enhanced the status of certain women in the chiefdom, as reflected in their membership of committees.

Changes are also evident in the direct relationship between the Chief's personal authority, commoner rights of usufruct, and political power. As the Chief's ability to give land diminishes, respect for his authority also diminishes, especially among the young and landless. Real Farmers influence decisions made by the Chief in the allocation of available marginal land and the determination of eligibility of people to get that land. They are also under using the land they have by usufruct, because they are able to support themselves through their involvement at Ntimane. All this means a radical redefinition of fundamental customary rights and obligations in the process of giving allegiance and gaining land. In practice, the land remaining is being selectively graded and allocated, candidates for land are being differentially graded, and proprietorial control, including rights of direct inheritance, is being sought over land currently used through usufruct.

Marginal land allocation

Marginal land is all that is left to give in Ngwempisi, because there has been a rapid increase in the number of dwellings over the last twenty to thirty years. In 1961, in the Ntimane sector, for example, there were sixty-one homesteads in clusters, concentrated predominantly along the lower and more fertile areas. By 1978, there were 107 homesteads in the same area (Government of Swaziland, 1961, 1979a).

In recent years the chance to farm commercially on Swazi Nation Land has become more widely recognized, and is now seen as a viable alternative to urban wage earning. Significantly, the status associated with urban wage work has diminished when contrasted with small scale farming enterprises in the rural areas. Many people expressed to me, with some disdain, their dislike of the need to work in the urban areas, and constantly referred back to rural areas, and rural living, as a cultural ideal. The result of this new heightened awareness of rural opportunity has influenced many young men to compete for land, even when that land is of poorer quality and dubious commercial value.

A new headman may have one of two major reasons for wishing to acquire marginal land. Either the headman is committed to a rural lifestyle and wishes to build a base in the chiefdom, or he is committed to working in the city for wages but wishes to retain a rural homestead. In each case, the headman measures the value of marginal land in very particular and subjective terms, entirely dependent on his individual circumstances and ambitions.

I found it necessary to distinguish between these two types of headmen, in terms of whether or not they are committed to living in the rural areas, and developing a rural based lifestyle. I chose the terms 'committed' and 'uncommitted' headmen to describe this distinction. Therefore, a committed headman is one who is determined to live in the rural areas, and is constantly developing his social and political profile. An uncommitted headman may choose to live in a rural area, or may do so by default, but shows little interest in the social, political or economic life of the area. These are not terms in use among Swazi, though all headmen in these two categories can be placed within the array of relevant social categories outlined in Chapter One.

A committed headman will be more concerned with the commercial viability of his marginal land than will an uncommitted headman, although this will be expressed in terms that do not relate directly to the fertility of that specific piece of land. For example, some other variables of importance to the committed headman may be proximity to water, distance from other arable land, access to roads, proximity to the homesteads of kin, and access to communal grazing land. In all cases, these variables will directly affect the ability of the committed headman and his family to maintain a rural lifestyle, although they will not restrict the wage earning capacity of an uncommitted headman. It is at this point that the influence of the Real Farmers becomes significant, in grading marginal land and influencing the allocation of that land.

The Real Farmers combine a farmer's understanding of fertility and yield potential with a traditionalist's intuitive understandings of the rights and obligations implicit in the use of land. Through their influence as members of the Land Management Committees, they are able to convince the Chief to grant specific pieces of land to particular headmen. Although headmen with strong farming credentials or ambitions may not be given much land, that which they receive is usually the land they desire for particular reasons. For example, both KaMotsa and KaGwebu are poorly sited and neither headman is committed to farming, intending rather that this land form the site of their rural based homesteads. Twala, Ntambo, Nhlabatsi, Mashinini, Mavimbela, Mkholo, Mavuso, and Mamba have all acquired small areas of reasonable land, well sited and conveniently located to water, roads and more arable land, giving these committed farmers further opportunities to expand their productive capacity. Mkholo, a particularly entrepreneurial farmer whose recent arrival was preceded by a good reputation as a committed farmer, was given a small, very rich piece of land immediately adjacent to the Mtimane, which he uses only for irrigating vegetables. Such is his success that he is actively sought after by Ntimane who want him as a member, thereby giving him the potential to increase his production further. One informant told me that Mkholo was favoured for membership because:

> He is very hard working and he is already successful. The farmers also like him because he makes a success of his own garden [his private garden] and will also be a success at Ntimane.

The worth of land can be further measured by a new headman's proximity to his kinsmen. Close proximity to kin allows ready access to an array of comprehensive infrastructural support, offsetting the disadvantages of poor quality land. In Chapter One I described the tools used in the homesteads, and the widespread use of oxen and single furrow ploughs. New homesteads usually do not have much agricultural equipment, and almost never have oxen, ploughs, or maize grinding equipment, because of the substantial capital outlay needed. They rely instead on family and neighbours for the loan of these items, and for assistance when they need it.

For example, KaNhlabatsi is the relatively new homestead of a rural wage earning committed headman who spends most of his time in the urban areas. His wife cultivates their small fields and, because Nhlabatsi's wage is low, she struggles to support her young children. She is constantly reliant on the assistance of other neighbouring homesteads for the use of their oxen at ploughing time and during the harvest. She has arrangements with one of her neighbours to use their oxen and sled, plus the assistance of a young man, to collect firewood. Further assistance comes from her husband's kinsmen in a nearby homestead, who provide her with seed grain, cattle pen manure and vegetables from their garden at Ntimane, in return for her help with planting, weeding and harvesting, and child care. The woman spends much time visiting her neighbours and her sociability is such that she can call for assistance when she needs it.

Access to grazing land can also affect the desirability of marginal land. All homesteads, whether on marginal land or not, theoretically enjoy equal access to grazing for their cattle. However, many informal arrangements actually govern access to grazing land. Headmen of marginal homesteads in the same locality as their fathers' homesteads are able to graze cattle over common land habitually grazed by the cattle of that lineage. Disputes related to grazing areas do occur, and tension results from the cattle of homesteads grazing over certain areas.

For example, KaMbingo has traditionally grazed cattle over Individual Tenure Land owned by an expatriate farmer. Recently, a Swazi businessman bought this land and refused Mbingo grazing rights. KaMbingo cattle are now grazing on Swazi Nation Land, causing some tension because they are grazing on land usually used by other homesteads. Since Mbingo is a long established headman with close kin in two other homesteads of the same name, he is able to gain the support of these headmen who allow him to use their grazing areas. In this circumstance a new homestead without close kin ties would experience some difficulty finding a solution.

As the above example indicates, a headman and his family can develop an array of local contacts and obligations that allow them to subsist. Once this has been achieved, a committed headman can decide to develop a more commercially viable enterprise based on that homestead, by asking for more land to cultivate, or by trying to enter an irrigation scheme. The right to ask for more land is a right allowable under customary law. Headmen have always exercized limited control over land, because the Chief allocated land on the basis of need. He did this based on the realization that, as the demographic configuration of the homestead changed, so too did the headman's entitlement to land. This procedure enabled a polygynous headman to accommodate additional wives and their children when needed, but was also premised on diminished rights to land as the configuration of the homestead changed. This traditional practice remains current in Ngwempisi.

An illustration of this process is KaShongwe, which had previously been a very large homestead with wide ranging fields. Following the death of the original headman, the other sons and their wives left, and the current headman, a monogamist, was required to give up some land to accommodate KaNhlabatsi. KaNhlabatsi, in turn, was required to accommodate KaMotsa, and KaTwala. In conjunction with KaShongwe, KaMotsa was placed under pressure to provide some fields for KaNhlabatsi and KaMagagula. In 1978, in conjunction with KaMsibi, KaMotsa also had to make way for the lower extension of the Ntimane gardens. The arrival of Twala, described above, a brother to the headman in KaTwala, caused quite some resentment, especially in KaNhlabatsi, because this heralded yet another claim on their land. The original homestead in that complex, KaShongwe, survived the re-allocations rather well by retaining adequate fields to the south west.

In the above example, only KaMotsa can be considered a marginal homestead with an uncommitted headman. Motsa's lack of commitment can be seen in his lack of emphasis on cultivation, and his reliance on cash income from urban work for subsistence. The tension between KaTwala and KaNhlabatsi is exacerbated by limited fields and a common commitment to farming. KaNhlabatsi and KaTwala are able to avoid competing for maize fields through their external cash incomes and their membership of Ntimane. KaMagagula relies heavily on cash income and some cultivation, with strong local contacts in Ntimane through KaShongwe and KaMamba. KaMotsa, although with limited fields, has membership of Ntimane. Significantly, the homesteads experiencing the least tension concerning land are either comparatively disinterested in farming, or have membership of Ntimane. Those experiencing tension have relatively restricted access to land and a commitment to farming, but no other local industry. In some cases, it is feasible that a headman can be committed not to rural farming but to rural residence, and can occupy a marginal homestead without any loss of status or opportunity, and thus without exacerbating tension. Ginindza is such a headman, who has developed a small service industry in addition to agricultural production. His cartage business supplements the subsistence production on his homestead, and he retains status and a significant profile in the local community. Since he makes limited claims on land he is able to avoid any tension with his neighbours over land.

The influence of the Real Farmers results in the most committed headmen receiving land with the most potential, according to a wide range of criteria, and based upon general acceptance of the Chief's right to re-allocate land according to need. Differentiation in rights of access to land has emerged in Ngwempisi at the most basic level, the process of acquiring the remaining marginal land in the chiefdom.

The Real Farmers have an important role in determining who should be eligible for giving allegiance, and what allocation of land they should have. The Real Farmers are well represented on the Land Management Committees by several of their most prominent representatives who strongly support rural opportunity for others such as themselves. These Land Management Committees are vital to the Chief, and receive his full support, because the land issues are constantly needing arbitration. For example, in the above mentioned re-allocations of land, it was the Land Management Committee for the particular area who decided the form that should take, and the Chief then ratified the proposal. The examples given illustrate two factors. First, land can be re-allocated according to needs largely determined by the Chief. Second, such allocations can have a very direct effect on the homesteads in question. In both cases, it is the Real Farmers, through their direct involvement as advisers and members of the Land Management Committees, who wield significant influence.

Restricting access to land

All young Swazi men have the obligation to give allegiance to a chief in return for land on which to build a homestead (see Chapter One). The implication of land scarcity in Ngwempisi is that many young men may have to give allegiance away from their father's chiefdom. The only part of Swaziland with available land is the unpopular, hot and malarial low veldt. Aspiring headmen do not want to give allegiance away from their father's area for a variety of reasons.

The move to another chiefdom entails spiritual, social, and economic dislocation, and is considered traumatic. The move sunders established kin relationships, and removes the family from proximity to the immediate ancestors. In relocating, a man is removed from the focal point of social and spiritual identity, as represented in the great hut at his father's homestead (Kuper, 1963, p.20).[5] A young man inherits many of his father's contacts, which are lost if he moves. As well, his father's chief knows him personally and he can win advantageous judgements regardless of his father's status. On the other hand, for some men a move to the low veldt may be advantageous, especially if they have poor reputations in their fathers' chiefdoms.

Giving allegiance in a new area entails building a new social profile without kin support. Economically, a new headman in a new area is seriously disadvantaged. In the early years of all new homesteads, returns are low, labour is scarce and financial outlays are needed, especially in the low veldt where conditions are not only more marginal, but infrastructures are comparatively rudimentary. These conditions affect cattle lending arrangements because cattle in the low veldt are more likely to die. Where labour is concerned, a young couple in a new homestead usually have the children of established kin stay with them to help with the work, in return for education. The relative scarcity of schools and clinics in the low veldt, compared to the middle and high veldt areas, makes parents reluctant to send their children to help relatives in the low veldt.

Except for the few aspiring headmen who may wish to avoid their fathers' chiefdoms, most aspirants are determined to give allegiance to their fathers' chiefs. This situation has ensured that becoming an eligible candidate to give allegiance is increasingly competitive, whereas previously it had been an unalienable right for all young men. I have concluded that, just as available land is being graded and granted selectively, so too are the recipients of land being graded. Their eligibility is considered by the chief in a manner that does not conform to customary norms of patrilineal descent and patrilocality as the main criteria for eligibility to give allegiance. Preference is going to individuals who bring more than just loyalty to the chiefdom. Prince Mkhosi, intent on encouraging initiative among his subjects, places a high priority on commercial farming and technological adaptation, and so emphasizes the productive potential of individual candidates.

Successful candidates for land, therefore, fall into three categories. Young local men with good reputations comprise the majority of successful candidates, and a good reputation is based on a range of criteria. For example, some young men are highly educated and have good employment prospects in the urban areas. Other young men have demonstrated their farming abilities on their parents' gardens, and are known to be industrious. The Chief told me that he considers the needs of young local men to be his first priority, and judges the worth of local candidates with an intimate knowledge of their kin connections, and usually some familiarity with their personal characteristics. The other two categories concern men who give allegiance from outside the chiefdom. To be successful, such men must be either preceded by a sizable reputation, as in the case of Ntambo, or have a significant local sponsor, as in the case of Twala. Ntambo was already well known to the Chief long before he asked to give allegiance. He had been working for a European farmer as a herdsman, and travelled constantly through Ngwempisi, in the course of which he had caught the Chief's eye. Twala came into the chiefdom from South Africa, through the sponsorship of his brother in KaTwala, who was already a prominent Real Person and whose wife is a Real Farmer at Ntimane.

In most cases candidates usually require some degree of sponsorship, and it is here that the Real Farmers wield significant influence. Young local men who have Real Farmers as kin, or have attracted the interest of significant Real Farmers, have a better chance of successfully giving allegiance. The Real Farmers of Ntimane have become significant sponsors of young local men. These young men are usually experienced farmers, often having worked in the schemes as labourers, for their parents or relatives, for most of their lives. Furthermore, by winning sponsorship of the Real Farmers they meet a demanding and highly selective standard, espouse the typical Real Farmer values of industry, and show a determination to cash crop. These standards reflect the stated preferences of the Chief.

For people born outside the chiefdom, sponsorship is also important and, again, it is based on Real Farmer values. For example, Ntambo told me that he deliberately sought to give allegiance in Ngwempisi because he knew the Chief favoured retainers who were hardworking and committed to commercial farming. In Ntambo's case, the Chief already knew him, but his eligibility was enhanced by the sponsorship of Real Farmers at Ntimane. I discovered that Ntambo was not only highly regarded by Ntimane Real Farmers before he arrived, but his older brother had given allegiance in Ngwempisi some years previously and, though now deceased, had been highly regarded.

Ntambo received a well sited, but very small, piece of land which was insufficient to support his homestead. He was then allowed to enter into various share farming arrangements at Ntimane, until such time as the development of section 'C' (see Figure 2:2) made it possible for him to become a full member with his own plot. Without sponsorship from Ntimane, he would have had to work in town to make a living, and would have had no way of proving his worth to the members of the scheme, a dilemma faced by many enthusiastic potential members. In the case of Ntambo, sponsorship from Ntimane certainly smoothed the way for his success. Aspiring Real Farmers who do not have Real Farmers as sponsors may encounter this lack as a limitation on their ability to achieve their goals. Ntambo, now one of the leading Real Farmers of Ngwempisi, bolsters the reproduction of Real Farmer values, in turn, by the considerable influence he wields as chairman of Ntimane, and as a leading adviser to the Chief.

To summarize the preceding points; allocation of remaining land, and determining who should be eligible to get that land in Ngwempisi, remains the responsibility of the Chief. He bases these decisions on direction from his advisers. In the case of land allocation, the advisers work through the Land Management Committees to determine how land should be allocated. They reach a decision by assessing a variety of factors relating to the long term intentions of the relevant headmen. The Real Farmers also influence the eligibility of men to give allegiance. They do so by fostering and sponsoring potential candidates through the irrigation schemes, confident in the knowledge that the Chief agrees with the need for agricultural intensification in Ngwempisi.

The previous sections of this chapter have dealt with the situation pertaining to land, when land is available. In Ngwempisi, land is scarce, and many young men already know that they will have difficulty giving allegiance to gain land for their own homesteads. I noted during my time in Ngwempisi that, while young men where concerned by this lack of opportunity, many were comforted by two alternatives available to them. First, young men accept that opportunities for personal gain exist in wage and salary earning in urban areas. Second, they increasingly look to their parents, rather than the Chief, for access to land and resources in the future.

This trend away from allegiance, to filiation in a conjugal family, is indicative of an even more fundamental change in the way people are perceiving their rights to land. This change is towards proprietorial control over land which they currently hold by rights of usufruct. This trend is powerful and pervasive, for the land in question constitutes the bulk of available land which has already been occupied for a considerable period under the house property complex. Headmen are setting out to acquire proprietorial rights so that in the future they may give the land to their children. Parents have expressed to me an interest in keeping their children with them and, when they reach adulthood, accommodating them on their homestead land.

Exercising proprietorial control

The consequence of granting proprietorial rights over land at Ntimane, while sustaining rights of usufruct at the homestead, has made many people aware of the distinction between the two places. Two major issues have emerged at Ntimane. First, the status of women has changed as opportunities have become available that were previously denied them. Second, plot owners are able to contemplate transferring land to their children, or whomever they wish, in a way not allowed under traditional rights of usufruct. Both issues are strongly related. Women are now vitally interested in commercial activity of all kinds in rural areas, and it is often women who wish to transfer assets to the children of their choice, rather than those chosen by their husbands, or determined by primogeniture. Real Farmers in general, are interested in long term commercial success. They have a high opinion of women's economic roles, and are keen to retain a reliable family work force in an increasingly commercial environment. They need to control the land.

It became clear to me that the Real Farmers are keenly aware of how they under use the land at their homesteads as opposed to Ntimane, and they are committed to gaining proprietorial rights over the land at their homesteads. I encountered difficulties in establishing this determination through conversation with the farmers. Such is the common awareness of land scarcity and the sensitivity of the issue that people rarely talk directly about their personal concerns and plans for land. Because land is an issue that rapidly arouses tension, people easily become jealous of each other, and jealousy leads to the distress of witchcraft. Instead, many farmers, while telling me that it was the concern of the Chief and the government to settle the problems of land scarcity, are in fact actively trying to guarantee that they keep as much land as possible for the future. They are aware that the pressures associated with land scarcity all over Swaziland are such that the government does intend to act at some time in the indefinite future to resolve the issue. Although at this point no definite proposals have been made by the government, I concluded that many Real Farmers are determined to be as advantageously placed as possible when that time arrives. To do this they must demonstrate that they need all the land at their disposal just to survive. The Real Farmers of Ntimane are the best placed to do this, because they can use the resources of their gardens at Ntimane to subsidize the systematic under use of their land at their homesteads.

Methods of under using and claiming land There are six distinct means by which homestead land is under used and controlled.

Labour usage at the homestead differs markedly from that used at Ntimane. Traditional labour roles are considered appropriate at the homestead, so most men have minimal involvement with homestead farming chores, which are done by women and children. With fcw cxccptions, men who hoe weeds at Ntimane do not do so at their homesteads. My informants make a distinction between hoeing in the homestead fields and the gardens by calling homestead labour 'women's work', a designation they never use for the same activity at Ntimane.

Distinctions in terminology and labour roles indicate particular social connotations evident in interactions at the homesteads. For example, *emalima* (work parties) are still organized at homesteads, although never at the irrigated gardens. Kuper stated:

> It is impossible for the homestead to satisfy its labour requirements at all times, and, apart from the assistance obtained from specialists, Swazi rely on sporadic communal work for a number of activities (1947, p.23).

In 1963, Kuper argued that the work party had a dual function; '... both [as] an economic venture and an occasion for the enhancement of prestige' (1963, p.46). People with irrigated gardens put much greater efforts into those gardens than into their homestead maize fields. Furthermore, the work parties I observed tended to contradict Kuper's economic rationalist description. For example, LaZulu the wife at KaShongwe, held a thirty-six hour work party over a two day period in September 1985, to complete a task requiring eight man hours of work. Over that period, this work party involved the labour of up to twenty people, who were clearly involved in a social activity which gave the host status. By contrast, cooperative work at Ntimane was managed efficiently, and governed by rules enforcing equal economic advantage for all members, with no clear status attributes beyond the coordinating role of the chairman.

A second distinction between homestead farming and that at Ntimane concerns farming logistics. Unlike Ntimane, where all gardens are tightly grouped, the maize fields are inconveniently and widely spaced, and separated by wide grass strips that prevent erosion.[6] Farmers at Ntimane expressed reluctance to use the communal seed plots because they were too far away from their gardens (a maximum of 100 metres from any garden). However, homestead maize fields are so widely spaced that the logistics of farming (spreading kraal manure and fertiliser, and carting the harvested crop) are expensive in terms of time and effort, and minimise the proportionate returns while maximizing the amount of area used.

Similarly, farmers who use only hybrid seed and chemical fertilisers at Ntimane use considerable quantities of traditional maize seed and kraal manure on their homestead maize fields. Some chemical fertilizers, hybrid grains and tractor ploughing are used, although not consistently, and only by some farmers. For example, Ntambo, who is a most progressive farmer, relies heavily on high yield inputs to achieve reasonable surplus production on his very small maize fields. Unlike at Ntimane, where there is a complete emphasis on high yield inputs, homestead maize fields are usually cropped with a variety of grains. Most of the commercially minded farmers adopt a semi-commercial approach to their maize fields, while non-commercial and semi-commercial farmers adopt a subsistence approach to maize production. They do this by adopting no risk farming techniques.

In 1979, a farm management survey conducted by the Department of Agriculture concluded that the use of traditional maize varieties was far more logical if the actual dynamics of local farming methods were taken into account. The cost effectiveness of local maize far outstripped that of the hybrid maize in circumstances where an average amount of land was available. Where access to land was drastically reduced, the growing of high yield hybrids made economic sense. Essentially, the survey concluded that farmers were either not prepared to, or could not justify, the cost of optimizing yield conditions for the hybrids. This was especially so since homestead labour is in increasingly short supply, owing to the large numbers of children attending school. The surplus of hybrids was found to be proportionate to labour invested in weeding, whereas production from traditional grains was unaffected by the presence of weeds. Since most weeding is done by women, it was concluded that women, in making economic decisions on the homesteads, were choosing the less labour intensive traditional maize over the higher yield, labour intensive hybrid grains, which did not guarantee a subsistence return (Government of Swaziland, 1979b, pp.ii-iii, 11-15).

Swazi farmers give priority to productive security over surplus production. This is well demonstrated by the selective manner in which some farmers use maize seed. The promise of a good season will be answered by the use of higher yield hybrid seeds and chemical fertilizers, in ground prepared well in advance by the tractor.[7] However, the main emphasis is always on guaranteed yields through the use of more traditional seeds and kraal manure. In a poor season, hybrids can yield a satisfactory subsistence crop if used in conjunction with traditional seeds, but at a much higher price. In good seasons, using combined inputs, there is a good surplus above subsistence needs, but without risk.

The 'no risk' option must reach a point at which it is no longer tenable using those techniques because of the limitations on available land. The farmers can only make a partial contribution to cash crop production by trading full efficiency for productive security. Eventually, as peer pressure, and economic and demographic pressures force greater commitment to cash cropping, farmers will more fully capitalise production. Under these conditions financial returns on capital investments will become critical. Currently, technological transfer is resisted by farmers who wish to understate the productive capacity of their homestead maize fields. However, because of growing pressure on available land, and the direct economic implications of using that land, it is inevitable that technological transfer must become more evident in the future, as farmers are forced to increase production.

Usable land is also being made non-productive by the use of four subsidiary techniques including wider grass strips, longer, narrower fields, widely spaced rows of maize, and finally, a situation where the most distant fields are cultivated first.

First, although the contoured maize fields have incorporated grass strips since the 1930s, the width of the grass strips has increased considerably over that time. When I measured old, unused maize fields I found that grass strips averaged less than one metre wide. In some old fields, the strips were as narrow as 300 mm. By comparison contemporary grass strips are usually in excess of several metres wide.

I assumed that the wider strips may have been more effective in preventing erosion. Further scrutiny of old fields indicated no discernible difference in erosion between these fields and fields currently in use. It is doubtful whether the grass strips even prevent erosion, except that they encourage farmers to cultivate along, rather than across the contour lines, a process that drastically reduces erosion in all parts of the world. Essentially, it is not the width of the grass strip that is important, but its contouring effect. Informants told me that the trend to wider grass strips was not a recent occurrence. From this I concluded that the widening of strips has been gradual process over a long period of time. Furthermore, this suggested that farmers using this technique have long been aware of the impending shortage of land.

Secondly, in the past, fields were narrower and concentrated more heavily towards the homestead, an arrangement that was logistically more efficient than the currently widespread fields. Increased productivity on maize fields has allowed farmers to use less total field area. Because farmers are trying to demonstrate their need for as much land as possible, they are spreading their fields out to fill the available space. Therefore, the maximum amount of land actually in use constitutes only a proportion of the land enclosed within the metaphorical or actual homestead boundaries. This can be seen in the way farmers arrange their fields to incorporate gullies and other waste land.

Thirdly, farmers also legitimately under use their expanded, widely spaced fields by planting maize in rows which are a generous distance (750 mm) apart. In densely farmed areas of Swaziland, the rows of maize are considerably closer together (300 mm).

One of the younger men with some formal agricultural training commented that he thinks land degeneration is avoided by this means of under use, as it negates the need to rotate fields. While this is feasible I concluded that a further factor lay in the farmer not having to reveal how little land could actually be used if the land was rotated. In one example, seven fields were planted with a total cultivated area of approximately one hectare (10,000 square metres). In each field were eleven or twelve rows of maize. These seven fields are situated in a total homestead area of approximately seven hectares, including the homestead itself. Work intensity was surprisingly low, involving two tractor hours (at E17.50 per tractor hour) and slightly over thirty hours of manual labour to complete.

Fourthly, farmers are also sensitive to the layout of their fields relative to other homesteads, and are concerned to place boundaries around the land they occupy. It is noticeable that fields furthest from the homesteads are cultivated first, despite the fact that cultivating the most distant fields involves more cartage and walking than would be involved in cultivating those closer to the homestead. This means that any land left unused lies close to the homestead, and well within any potentially disputed area. This technique, like the technique of widely spacing rows, ensures that all available space is used. The result is that the farmers use the maximum amount of area, yet achieve only modest production. The farmers must be seen to utilize all the land to make a subsistence return. Otherwise they would be demonstrating that they could use less land to achieve the same result.

Fifthly, it is feasible that the previous four mechanisms for under utilizing land may be a genuinely subconscious response to land scarcity by the farmers. However, a proliferation of fences constitutes a conscious and premeditated attempt to delineate private boundaries. Fences are a potent symbol of changing attitudes to land tenure in Ngwempisi. Some fences are strong and comprehensive, while others are weak and shabby, but all indicate some degree of proprietorial intent over the land.

Three types of fence are evident. Fences bounding the irrigation schemes are strong and secure. In contrast, homestead fences are generally inferior and of two types. Close to the homestead, the fencing is of a reasonable quality and can be made secure. On the peripheries, fences tend to be of very poor quality, often to the point of having no wire.

The strong, secure fences surrounding Ntimane and the other irrigation schemes are a definite statement of tenure. Built under the auspices of the Rural Development Areas Programme, these fences clearly delineate the intensive farming areas from cattle grazing areas, which were set up by the Rural Development Areas Programme to provide areas where unsupervised cattle could graze. The fences also delineate settlement areas and are well maintained. The areas of the irrigation schemes are neither in dispute nor available for re-allocation. For example, at Ntimane, tenure has been determined and defined; the fence protects the crops from animals and symbolises definitively the tenurial status of the farmers. Gates at Ntimane are closed, unlike those at the homesteads. By further restricting the access to land with closed gates, farmers make a more definitive proprietorial statement. On the other hand, when gates are left open and the passage of stock is allowed, some acknowledgement is made of limited rights over common ground. Where gates are always closed no presumption of common land rights can be assumed by other people.

The variety of poor quality fencing evident around the homesteads is indicative of the uncertainty the farmers feel over their homestead land, and of their attempts to lay claim to that land by delineating boundaries. These fences usually mark a line between the homestead of the fence builder and his nearest neighbour. Often without wire, the fences allow the free passage of stock; in fact, cattle and goats are often deliberately driven through the fence by way of a short cut to the cattle pen. Many fences begin and end in an arbitrary manner, without actually enclosing a piece of land.

The farmers claim that they need fences because their sons, who would normally herd cattle, are now attending school. However, all cattle are still herded, either by boys of the family, or Mozambiquan refugees.[8] I saw clear evidence that the cattle were closely supervised whenever there was a thunderstorm. On these occasions, the herdboys came running with the cattle back to the cattle pen. The primary purpose of returning is for the boys to seek shelter for fear of lightning strikes, but the cattle are also returned to prevent them from entering maize fields without supervision and eating the young maize shoots. The argument that fences have replaced herdboys is clearly erroneous, although, admittedly, herdboys have a busier, lonelier life than in the past, when labour of this kind was more abundant.[9]

When queried, farmers were bemused by the poor quality of fences, and were unembarrassed by the contrast between the strong fences at Ntimane and the poor homestead fences. Yet the farmers were reticent to elaborate on why they persist in building these fences. I gained a major clue to their fence building behaviour when a prominent Real Farmer told me that he built a twenty metre fence across the end of his maize fields to restrict the activities of another farmer who was slowly lengthening his fields onto the Real Farmer's land.

When fence lines are plotted on an ortho-photographic map of the area, it can be seen that many fences are obviously open ended, i.e. they do not entirely enclose any piece of land and cannot be considered stock proof fences. By building such fences, the farmers apparently are attempting to delineate the boundaries of their homesteads and fields from those of their neighbours, and from footpaths and tracks that they feel begin to encroach on their land. For example, Msibi, who has not fenced his homestead, has nevertheless fenced off the field that borders a nearby track. Initially, a long fence delineating a grazing area ran the length of his own and three other blocks. Residents of other homesteads gained access to their homesteads by passing between Msibi's field and those of the neighbouring homestead. During the wet season, motor vehicles trying to avoid mud holes deviated into Msibi's fields. In response, he placed posts and several strands of wire along the road edge. Any deviations that occurred subsequently were forced away from his fields towards the other homestead. Msibi's intent was not stock control, but traffic control, and was immediately effective in ensuring a boundary for his property. In another example, a homestead was fenced in such a way that the pit toilet ended up located some distance outside the wire. Because of the placement of the nearby road, the headman claimed the maximum amount of land by leaving the toilet outside the delineated area.

Farmers are using gullies and waste land in determining boundaries. This unusable land will also provide future homestead sites, leaving arable land free. One farmer fenced a large area with high quality rabbit wire, enclosing an area of scrub and sheet rock. He is interested in water catchment and in providing land for his sons to live on, envisaging that this non-arable land is where they will place their homesteads.

There are two reasons for the difference between the fences near the homestead which are strong and practical, and those on the peripheries which are shabby and ineffective. Headmen feel confident about their control of land close to their homesteads, hence stronger fences are erected at the centre. As they move towards the boundaries of their properties, they are careful to avoid antagonizing their neighbours, so they try to mark off an area without making a strong tenure statement. In time, once their legitimate occupation of the peripheral land has been confirmed in the minds of their neighbours, they will build stronger fences at the peripheries.

The second reason for the variable quality of the fences relates to the difference between headmen committed to monogamy and those who are committed to polygyny. The monogamists have the strongest fences because they recognize that, according to tradition, they are entitled to enough land for the one wife. By building strong fences the monogamists are making a strong tenure statement. Polygynists, on the other hand, are afraid to delineate their claims too clearly by fence building, in case it negates their traditional rights to claim more land in the future if they get more wives. They tend to build weak fences as boundary markers to stop encroachment, rather than fences delineating total homestead area.

There is a noticeable relationship between strong homestead fences and solid permanent buildings. This was confirmed by an Rural Development Area fieldworker who argued that once people have individual tenure (through resettlement) '... immediately they build solid [concrete] block houses'. Concrete block housing, rather than mud and stone, has become a favoured form of construction since Cyclone Domoina destroyed many mud huts in 1983. Farmers cite the strength of concrete as the reason for building permanent dwellings, and so react against suggestions that their heirs would need to move the homestead after their death.[10] After such a considerable effort expended building concrete block huts and houses, the headmen intend those exact sites to continue to be occupied. Zulu, for example, is an aging headman whose son has built a multi-roomed house of considerable size on his father's homestead site. It is quite clear that his heir intends to occupy this exact site after Zulu's death. Wherever there are homesteads of this type, strong fences are usually also evident.

By means of these techniques for under using, and ultimately claiming proprietorial rights over land, headmen and their wives ensure that, in any resolution of the land tenure issue, they can convince the authorities of the full extent of their need. Among Real Farmers, there is a complementary trend to monogamous marriage, a practice now encouraged by christian beliefs. Polygynous marriage is losing its appeal as the neighbours of polygynists react against the additional claims polygynists make on surrounding land. Polygynous marriage is further marked by female antipathy to the institution of marriage, particularly when women, being aware of land scarcity, are reluctant to marry a man who cannot guarantee them their own house and fields.

In monogamous marriages, the inevitability of a conflict of interest between competing wives and their children is eliminated. The economic interests of men match those of their wives, as the lack of counteracting claims on the man forces a new relationship between a father and his sons, and between the sons and their mother. The mother's traditional role, as intermediary on behalf of her eldest son's interest, is subordinated. This represents a basic, although currently incipient, point of division between the customary interests of mothers and their sons. However, the relationship between fathers and sons becomes more directly linked, and carries with it the idea of direct inheritance from the father. For example, one headman stated:

> We should make things like villages. In this case I mean that my sons shouldn't go out and *khonta* [give allegiance] but they should stay all about my homestead and reduce this overcrowding. Now everyone has a big homestead covering a big space ... when the land runs out we could build up, the father below and the son on top! [laughter] Swaziland is small and if people want to farm and keep cattle in a small area it is impossible, so they must occupy a small space'.

This statement from a prominent Real Farmer could only come from a monogamist, as indeed is the case. What is interesting is that the man is assuming that he is responsible for providing land for his children, when this has always been the Chief's responsibility. Not all headmen are as clear about their intentions as this farmer, and many indicated to me that they are confused about their children's future. Under utilizing homestead land may be a deliberate strategy by some farseeing farmers; for most, however, it constitutes a resort to conservative cautious behaviour, in a future that is increasingly uncertain.

These major changes in definitions of tenure are increasingly supported by a strong moral imperative, which is based on zionist christian influences in the promotion of monogamy and radical re-definitions of customary notions of social space. Accepted understandings of the homestead as a 'site' and a symbol, and an anchor point in Swazi social relations, are changing rapidly. The change is away from the idea of individual control over land as an alien concept, to one where social space is being actively extended and objectively measured. This process of extending social space involves rejecting customary obligations to give up land that is no longer needed by the homestead. It also forces people to enter the politics of space on a broader scale, focusing interest on the conjugal family, as located within a specific space, at the expense of the interests of the lineage within the general chiefdom.

The Real Farmers lead by example in this process of under utilizing homestead land. To do this they must be able to offset lost income from homestead fields against income from their gardens at Ntimane. But they must also assuage the feelings of a variety of parties. First, they must demonstrate convincingly the full extent of their land needs to their neighbours and to the Chief, in order to avoid jealousy and suspicion on the part of their neighbours. Second, they must maintain the size of their holdings until the government, which is moving very slowly on the land tenure issue, passes legislation that re-defines rights in land. Finally, they are trying to control the Chief's urge to concentrate more people on his land. Therefore, the farmers manipulate public perceptions of their control of space by reconciling a contradiction. They must prove that they use land responsibly, while demonstrating that they need as much land as possible. In this situation farmers must convince everyone that they are innovative and deserving of land, while they deliberately minimize production on land to which they have no title.[11] The Real Farmers must force a continuing distinction between tenure of homestead land and the land at Ntimane, and in doing so must dichotomize the two institutions. If they fail to maintain the dichotomy in the minds of their neighbours, then they cannot justify their lack of intensification on their homestead fields.

Even more complex is the fact that Real Farmers' techniques for under utilizing homestead maize fields are in direct contradiction to the Chief's interest in accommodating more retainers. The Real Farmers must be able to convince the Chief that they act in his best interests, and still retain strong traditional imperatives, through membership of traditional decision making bodies, while systematically under utilizing their homestead maize fields. If they are successful, the Chief's ability to accommodate more retainers is severely limited.

The three case studies which follow illustrate the differing abilities of people to cope and adapt in Ngwempisi, and the sorts of choices they make depending on their commitment to monogamy and/or to commercial farming.

Case studies

Case study one

In this homestead, the headman was an early member of Ntimane, but is not seen by the other members as a Real Farmer. He became a member of Ntimane after losing his maize fields to make way for the original gardens. He has two sons, one by a wife who is now dead, and he is a traditionalist non-christian. Known as a sour, taciturn man, he has failed to utilize his irrigated garden in the way the Ntimane members expect, causing vocal, although muted, annoyance and resentment among them. Criticism of his lack of commitment to commercial farming has been carried over into other issues. For example, he is seen as reactionary because he has not developed either his garden or his homestead, which are both of poor quality. He also does not use the Ntimane equipment, and this is interpreted by the other members as indicating an uncooperative attitude.

Despite his membership of Ntimane, this headman is becoming marginalized. He is disadvantaged by his lack of motivation to be a commercial farmer, and is increasingly out of step with Real Farmer labour ethics. Disapproval of him has not been moderated by his traditionalist habits, evident in both his mode of dress and lifestyle. He was identified, along with six others, as a wizard at an *umkhaya* (witch hunt) in the late 1970s, an accusation that satisfied many of his critics, and provided justifications for their suspicions. A slight propensity for anti-social behaviour, and his obvious reluctance to be a Real Farmer, are interpreted as apathy and jealousy. For this man, the social situation he finds himself in is now beyond his control. Early pressures to conform to Real Farmer ethics have become overwhelming as the Real Farmers have grown in moral stature and gained in popular support.

The inevitability of this headman's marginal status was suddenly shattered in December 1985 when, as he walked to Mliba, he was 'accidentally' killed by a boy who threw a stone at him.[12] It was agreed among members at Ntimane that, after her mourning period, his widow could have tenure of her husband's neglected garden, which she would work with her son. This woman is the deceased's second wife and her son is not his official heir. However, as his heir by primogeniture is mentally handicapped and therefore incapable of membership at Ntimane, the members are pleased with the change of circumstances that allow the well regarded widow to re-establish the garden.

The headman's unwillingness to adopt an innovative approach clearly prejudiced him in the minds of the Real Farmers and the other members of Ntimane. Intolerance for his lifestyle, ultimately resulting in an accusation of witchcraft against him, is a potent indication of how Real Farmer ethics have permeated the social fabric of Ngwempisi. Conscious social sanctions were invoked against this man to attempt to enforce particular standards which are determined by the Real Farmers.

Case study two

This homestead is an example of the classic hybrid modern/traditional model of female economic management, with a cash earning absentee headman, and a nucleated and monogamous family situation. The head of this homestead is not an Real Farmer, but is a headman and lesser Real Person. The organized force in the homestead is his comparatively young, but immensely popular, second wife, who is a prominent Real Farmer. This woman has six children, five of whom are boys, and a reputation in the general community for hard work. The homestead head was one of the early members of Ntimane, and has always had a garden there. A prominent Real Farmer commented that it was solely due to the organizational ability of the woman that the homestead has become so well equipped. At Ntimane, she is so popular that the garden is considered to be hers, despite her husband's claim to the contrary.

I was told that, prior to her marriage to the headman, the homestead was becoming marginalized. Such was her energy in re-organization that, in reversing the trend, she made enemies among her husband's relatives, which resulted in witchcraft accusations against her. Her response was a public challenge to her accusers who immediately retracted the accusations. Her courage and forcefulness in the situation enhanced her status. She is the dominant personality in the homestead and embodies all the ethics of hard work and managerial skill so respected by Real Farmers. The witchcraft accusations are still commented on, and laughed about, by various members of Ntimane.

The homestead has a good cash flow because of the headman's relatively well paid job. He is home briefly once a week, and so all management decisions are made by his wife. Under her direction, the homestead buildings are being upgraded to concrete block, fences are gradually being built, and the fields are comprehensive and well cultivated. The woman has a high social profile, enhanced by her involvement in social activities outside Ntimane. For example, she has co-ordinated work parties and is a member of the Cattle Dip Committee.

Her control in the homestead is comprehensive. On one occasion she called a meeting of her husband's male agnates to arrange compensation from them to her for the death of a cow accidentally killed by their common herd boy, who was her husband's male relative. Compensation was arranged and given, the cow was replaced, and the matter closed. In a number of ways this unusual happening demonstrated the degree to which roles have changed. First, the cow belonged to a male of her lineage who had given it to her under a farming out arrangement, and was one of several cows and a herd of goats that she controlled under such an arrangement. Theoretically, the high degree of her involvement with livestock should not be possible, for husbandry of larger animals, especially cattle, is a man's vocation. Second, she conducted this meeting without the significant involvement of her husband, even though senior male kinsmen of her husband's were present, as was a prominent, although unrelated, Real Farmer, who acted in an adjudicative role.

In this homestead little sex based labour differentiation occurs. Resident boys, although not all her children, do manual labour associated with women's work, and the homestead's herd of cattle is often looked after by a small boy or girl from a neighbouring homestead. This woman is active at Ntimane, and enjoys the public support of prominent Real Farmers. On several occasions she has been publicly complimented for various leading roles she has played in Ntimane business.

This homestead is increasingly competitive in the current economic situation. Even with under utilization of the maize fields, a surplus of maize is produced for sale, raising about E400-E500 per annum. The homestead exemplifies the domestic mode of production, but through the woman's involvement at Ntimane, a strong commitment has been made to providing an alternative cash income to that earned by the headman in his salaried work.

Case study three

This homestead belongs to one of the most prominent of the Real Farmers, who is the influential founding member of Ntimane. This man is also one of the leading Real People in Ngwempisi. For many years now, his homestead has been fully fenced with strong cattle proof fencing and secure gates. The design of the homestead suggests a monogamous intent in marriage which is practised within a christian tradition. This farmer has been committed to cash cropping on Swazi Nation Land for many years, and instigated cash cropping in the area.

Unlike most of Ntimane Real Farmers, this farmer is prepared to carry out some intensive maize growing at his homestead. I believe that this indicates just how sure he is of the security of his position in the chiefdom. He is able to use his homestead land in a proprietorial manner without feeling that his control over the land is threatened. In this, he is the exception, as I found no other Real Farmer who had a similar security of tenure.

The farmer's primary cash emphasis, however, lies in vegetable growing at Ntimane, where he employs labour to help him manage two gardens. Contracts for vegetables are held with a local secondary boarding school in Mliba, and with suppliers from as far away as South Africa. External cash income is earned by one son, as yet unmarried, who works in the low veldt, although this income may not necessarily accrue to the homestead.[13]

This man is a member of the Land Advisory Committee, and has great influence over recommendations concerning land tenure that are made to the Chief, though this has not resulted in excessive acquisition of land on his part. However, he has been able to make arrangements over the use of land in the past, in a manner not usual for a commoner. For example, I was told by his son-in-law that his father-in-law had given his daughter some land some years ago, and that there had been no repercussions.

In another instance, the same son-in-law was given some land by his maternal grandmother[14]. Male relatives of the maternal grandmother made claims on the land and the son-in-law moved to land made available by his father-in-law, the Real Farmer. Once again the Real Farmer demonstrated his ability to make land tenure arrangements on his own initiative.

It seemed that, since the Real Farmer could fence his land and give it away with impunity, he must have the strong support of the Chief. Subsequently, in conversation with the Chief, I discovered that he did indeed trust the Real Farmer's judgement and allowed him much leeway in matters of this kind.

This homestead is representative of those which have best adapted to change in Ngwempisi. One son appears ready to inherit, and he already has tenure of a garden which his father uses. He has been educated, and on his father's death will acquire the homestead, maize fields and rights over two gardens. Significantly, this son is also the youngest son. This pragmatic selection of the youngest and most suited represents the best of selection procedure in a polygynous situation, but is unusual in a christian monogamous situation where primogeniture is the norm. The other sons have all been educated and work in urban areas.

This headman is widely recognized and is a gifted public speaker with proven leadership abilities. His public persona is charismatic and inspirational, and he is able to convince people that his actions are legitimate and conform to tradition. Privately, he is politically skilful and knowledgeable, and fully aware of the long term implications of land scarcity. With the support of the Chief, and through his support of the chief, he has systematically worked, over a long period of time, to achieve proprietorial rights over land both at Ntimane and at his homestead.

Conclusion

In this chapter I have argued that the Real Farmers are seeking proprietorial control over land which they currently hold by rights of usufruct, and that this process is resulting in increased social differentiation. This process has three aspects. In the first, the Real Farmers influence the distribution of the marginal land still available. In the second, they influence the selection process by which candidates gain access to that land. I have presented data suggesting that by these means the Real Farmers restrict access to resources to those who espouse similar values to their own.

The third part of the process involves Real Farmers systematically under utilizing homestead land in an attempt to acquire proprietorial rights over that land. It was clear that the Real Farmers can only sustain their under utilization of homestead land because they are members of Ntimane, and also because they have some influence over the decisions the chief makes, through their representatives on the Land Management Committees, and among the Chief's advisers.

The processes by which the Real Farmers gain land are exclusive of increasing numbers of people. Social differentiation is becoming more marked and inequality of access to land more pronounced. Many young Swazi will never get land. The children of unmarried women, of urban based wage earners, and of men who never gave allegiance, are numerically significant categories of people, one degree removed from rights in land.

Even within the chiefdom, among those who have land, the tempo of competition is further differentiating commoners. Some headmen and their families are adapting well to social change, while others who are not are becoming economically and politically disadvantaged in the area. The headmen who are doing well, such as the Real Farmers, demonstrate an ability to adapt to the demands of land scarcity, and have a strong role in the political processes that are evolving in Ngwempisi.

Notes

1. Swaziland has a high and accelerating birthrate, projected in 1983, at 3.41 per cent per annum. Of this population only about fifteen per cent reside in urban areas (Government of Swaziland, 1985, p.33).
2. Estimates of soil loss run to 3,000 tonnes of soil per square kilometre per year (Times of Swaziland, 28/8/85). These estimates were made in the press by Dr. Mike Wagner, Head of the Department of Geography at the University of Swaziland and he later confirmed this in personal communication with me.

3 The KaNgwane Land dispute involves areas of South Africa to the south east and north west of Swaziland, and sovereignty over 800,000 ethnic Swazi, with attendant refugee problems (Billard, 1986; Government of Swaziland, 1982b, 1983; United Nations General Assembly, 1984).

4. Ritual murder is the killing of a person to gain human parts to make powerful medicines for personal enhancement.

5. Kuper (1947, 1963) and Marwick (1966) state that the relationship with the land is strong, and has an intensely spiritual aspect. The kinship based tie to the land is considered significant, so that any lapse in tenure will see the land resumed by the chief, but subject to claims by the relatives of the former tenant (Kuper, 1947, p.46, Marwick, 1966, p.161). Men are also drawn to the land on which the graves of their ancestors are found.

6. The grass strips following the contour lines of the land were the result of Sobhuza II deciding in the 1930s that erosion needed to be combated and implementing the grass strip system. Grass strips are now quite distinctive features of Swaziland's rural areas.

7. If it is possible, early cultivation is considered to be important. In the past this was limited by the fact that oxen could not plough the hard ground, now the tractor has overcome that problem.

8. Boys tend to be rotated through school, working two years at home and then two years at school. Mozambiquan refugees are in the area in large numbers; there were forty in the Ntimane sector of Ngwempisi.

9. Older men say that in their childhood herding was fun because lots of boys would play together in large groups. These days the boys find it a to be a lonely chore, with many more cattle to look after, as they might have the cattle of several homesteads under their care.

10. According to tradition, after the death of a headman a homestead should be moved to another site (Kuper, 1963, p.21).

11. Some farmers claim that this situation will foster corruption, through people with money pressuring the chief, through his *Indvuna*, (Governor) for land. However, most people with whom I discussed the issue said that corruption was not yet an issue in Ngwempisi.

12. The boy had been frightened to see the man, because of his status as a wizard, and had thrown the stone at him possibly to frighten him. Instead the stone struck him and he died.

13. The importance of remittances to the homestead is well established in Swaziland as in South Africa. However in this particular situation, where the headman is well financed, it would not necessarily be a subsistence requirement but may be managed in the form of cattle purchase and cattle lending arrangements.
14. This is another example of an occurrence that should not be possible, but can occur if a very forceful woman wishes to give something to younger sons or grandsons who are not heirs.

4 Labour and finance

Introduction

In this Chapter I will examine labour and finance, and the manner in which the Real Farmers uphold tradition while encouraging other non-customary practices. Specifically, I will illustrate how, by differentiating between their homesteads and Ntimane in labour and exchange practices, the Real Farmers are beginning to dominate the economy of Ngwempisi.

Labour

Labour is narrowly defined at Ntimane, where members distinguish between community work and individual work. There is no division of labour by gender, members are expected to work their plots, and people are hired as agricultural labourers. Alternatively, at the homestead, different criteria exist. Work patterns follow the traditional model, a division of labour by gender remains dominant, absentee homestead heads are common, and labour is commonly kin based and does not involve fixed wages paid directly to the labourer.

Homestead labour patterns

Work patterns at the homestead are different from those at Ntimane. The most basic divisions are those of sex and age. On the homesteads, as a general rule, labour tends to be predominantly that of women and children. This is a result of a general awareness that to adopt similar work patterns at both places implies no difference between the homestead and Ntimane. Farmers are also aware that their ability to under use homestead land is financed by their gardens, so if they were then to draw labour away from the gardens, this would jeopardize their strategy of under using their homestead land.

There are men in the Real Farmer category who, in contradiction of this general rule, labour in their homestead fields as people do at Ntimane, but these men are usually non-member Real Farmers, and many are commercial maize producers. A very small category of producers, they are adopting non-traditional labour patterns at their homesteads so that they can demonstrate rights over homestead land through intensive production. They believe they are justified in these actions since they have no commercial alternative, as do the Ntimane farmers (see Chapter One).

This chapter concerns those Real Farmers who, wishing to sustain the under use of homestead land, put their own labour into their commercial gardens, but utilize traditional labour patterns for homestead subsistence production.

Some caution must be exercised in extrapolating on the basis of the evidence of male labour at the homestead, for in many instances a lack of male commitment to labour at the homestead is offset by strong female management. In some cases, this could be considered part of a general strategy, remembering that many women can be called Real Farmers (see Chapter One). There are powerful incentives for women to dominate homestead labour for, despite some land rights through their membership of schemes, membership of homesteads enables women to gain political influence and social status. As previously described in Chapter One, it is through the headman that official political representation is gained. Women can gain in wealth and power, but it is when they do so through their sons that they are most powerful, since they exercise great control over their sons and their families. Even Real Farmer women know that, in terms of tradition, they are in a stronger position if they operate though the headman, either pliant husbands, or their sons.

In terms of the house property complex, by which a woman invests in her son's potential, she has very direct interests in the continuance of the homestead and especially in her domination of that environment. For example, Shongwe is a frequently absent salary-earner who limits his labour involvement at his homestead to tasks such as using oxen to weed which are usually carried out by men. In the 1985/86 planting season Shongwe worked on one occasion, and the work was organized by his Real Farmer wife LaZulu, a strong and opportunistic manager, who uses boys and young men for many tasks traditionally associated with women's work.

LaZulu does all the work at Ntimane for KaShongwe, with limited assistance from her children. At the homestead she draws in external labour under informal arrangements, and except for some particularly heavy periods, external labour is the only labour used about that homestead.[1]

The work at KaShongwe involves two young men, one of whom is resident at the homestead, while the other lives some distance away. In both cases, the young men are fatherless, and therefore have few good prospects for the future. The former is being sponsored in his education by LaZulu, while the latter is paid a direct wage. LaZulu's intention, in using the young men at the homestead, is not to create personal ties with them, but to maintain close contact with their families.

In the first example, the youth's mother has married a male kinsman of Shongwe's, and the youth is considered a potential drain on the new husband's resources, especially since the woman has children by that man. By allowing her son to live at KaShongwe, and by exchanging his labour for his education, both the youth's mother and her husband's kinsmen are pleased that he can live and work at KaShongwe. This situation results in an obligatory relationship between Shongwe and LaZulu, and Shongwe's kinsmen.

The second young man has never been fully filiated within his father's lineage. His biological father committed a capital crime, and his mother returned to her father's homestead. This headman is a senior member of Shongwe's lineage, and was obliged to accept back his daughter and her son, as few marriage cattle were ever exchanged.

Technically, the young man is still a member of his mother's father's lineage, and therefore he could only give allegiance in Ngwempisi. As a result, he needs to acquire social standing in the area. Unfortunately, the headman has a large homestead and is heavily financially committed, and so the young man and his mother are a considerable drain on resources. LaZulu uses this young man's labour, and pays his wage, of E50 per month, directly to his grandfather. This allows her to satisfy her own labour needs at KaShongwe. It also satisfies the interests of the young man's grandfather in a variety of ways. The arrangement provides a cash income to the grandfather's homestead, gives the young man a labour profile in the area, and provides him with future employment opportunities. The grandfather thus has his grandson permanently resident at his homestead, can use his labour as he needs it, but does not need to support him. In this instance, LaZulu has additional labour as she needs it, while building further strong linkages, through a network of kinship obligations, with her husband's kinsmen.

LaZulu's active managerial role relates closely to her ability to get access to funds, either through her garden at Ntimane, or from her salary earning husband. By dominating the economic environment of the homestead she ensures her own long term interests, because ultimately her status and social position depend upon the succession of her son. She also directs the homestead's resources into her children's education, and so ensures a spread of potential for the future.

LaZulu has a variety of needs. She must place emphasis on her children having a profile at Ntimane, where she controls inheritance and can influence the selection of members. The children are also freed from chores, by her labour and the other labour she uses at the homestead, to enable them to receive a full time education while they are still young. At the homestead, she requires more labour than she alone can provide, and so calls upon arrangements made through her husband's agnates to gain access to labour. She thus simultaneously enhances her house's prospects in a direct commercial way, whilst enhancing her son's (the future headman) profile among his kin, by building a significant social profile among her son's lineage elders.

Labour at Ntimane

The use of kin based labour at the homesteads, as illustrated by the example of KaShongwe, is increasingly contrasted at Ntimane by the use of labour that incurs few accompanying social obligations. This trend is most obvious in the use of non-kin contract labour, an arrangement which involves little reciprocity and is aimed at reinforcing kinship ties, by avoiding client to patron relationships emerging between kinsmen.

Farmers began to employ contract labourers because there are many *Shangaan* (Mozambiquan refugees) in Ngwempisi.[2] The refugee men are sometimes employed to work on irrigated gardens, often work as herdboys, and in one case a *Shangaan* was employed to manage a garden. The use of

Mozambiquans, coming at a time when there was a ready supply of these young men in desperate need of support, facilitated the rapid introduction of wage labour into Ntimane.

Noticeable differences are immediately apparent in attitudes to *Shangaan* in employment, as opposed to Swazi males. Swazi wage labour is disguised as part of a kinship obligation, and is justified by the wages being paid directly to the employee's parents. In this way, the Swazi employee is never paid directly, and is rarely directly criticized for poor performance. The *Shangaan*, on the other hand, are paid directly and are also openly criticized. They are often accused of being sly, duplicitous, and thieving. Despite this vilification, the *Shangaan* continue to be reasonably well treated, and seem to be fairly happy with their lot.

This ambiguity in attitude can best be explained as employers playing down their preference for contract labour over kin based labour. By vilifying the *Shangaan*, an employer is able to reassure kin that work opportunities for young Swazi are not threatened by the presence of the *Shangaan* contract labourers. Further controls on the use of contract labour exist. For example, the employer who allowed his *Shangaan* worker to manage his garden unsupervised attracted some disapproval from Real Farmers, who also severely criticized the quality of the *Shangaan*'s work. The *Shangaan* is not employed in this way any more. By allowing the *Shangaan* to take on too much responsibility, the employer was perceived to be undermining local confidence in kin based labour structures.

The *Shangaan* would become more acceptable if they were eligible to give allegiance, but this is not possible because of their alien status. Since they are effectively excluded from any opportunity to gain status or land, young Swazi women consider them to be poor prospects, and do not seriously consider them as lovers, despite their reputation as flirtatious. Young *Shangaan* have also acquired a reputation for being sexist in their attitudes to labour, which further prejudices them in the eyes of the Real Farmers. For example, in one homestead, a *Shangaan* was evicted by the female manager for refusing to do what he described as 'women's work'. The reaction from among Ntimane members was to praise the woman for her progressive and appropriately flexible attitude to labour.

Since the *Shangaan* cannot change their status and are thus forced to continue in contract labour, the situation suits the contract labour needs of the

Real Farmers perfectly, and allows them to get their own children through school much more quickly.

At Ntimane, a noticeable change in labour roles has occurred, since men have more readily acknowledged the value of female labour. Though men may avoid adopting women's work at their homesteads, they do assume what are otherwise defined as female labour roles at Ntimane. This indicates a fundamental change in attitudes to certain kinds of work by men, and is a feature of the Real Farmers. The most dramatic consequence of this changing attitude is in the status of women.

Generally, in Swaziland, women are considered more economically minded than men, and more capable of generating a regular income. On the other hand, men are considered to have fulfilled their familial obligations if they sporadically remit money home. The situation is markedly different among the Ntimane members, where women are valued members. Currently fifteen of the forty-five plots are owned solely by women, and many more are worked exclusively by women in circumstances where the male plot owners work in the city. In these instances, the wives are treated by the members of Ntimane as the owners of the plots. Circumstances such as these mask the fact that women constitute a large part of the commercial enterprise at Ntimane, to a degree not indicated by formal ownership. I was informed that the comparatively large number of plots registered to absentee males stemmed from an initial reluctance to register women as members because of reservations about the legality of such an action. As a result, many plots are registered in the names of males, but are actually recognized as belonging to women.

The Real Farmers place a strong emphasis on qualities which they expressed as 'honesty', and an ability to 'work hard'; for the most part such qualities are believed to be peculiarly female qualities, and essential to the success of Ntimane. Not only are these female qualities upheld as being appropriate for male members of Ntimane to emulate, but they are reflected in the manner in which formal policy and planning are directed at women. In his addresses, the chairman of Ntimane always directs his attention quite specifically to women, who all sit together to one side of the shed. Women receive particular recognition for their ability to act in a cooperative manner, which is considered an advantage in organizing communal labour.

The adoption of women's work by male Real Farmers has had a significant effect on administration at Ntimane, and a direct effect on Real

Farmer control over labour. Administrative influence takes the form of direct and formal involvement in organizational activities. As I have previously noted, four of the seven committee members are women, and this trend has carried over into positions of authority in new developments outside the scheme.

The prominence of women at Ntimane is not necessarily reflected in positions of real power. While women are numerically dominant on the committee, males occupy the roles of chairman and vice-chairman, though outside the scheme a local woman is chairman of the Cattle Dip Committee,[3] while another heads the Pipeline Committee. The distinct seating positions of men and women during community meetings demonstrate direct status attributes, as all prominent positions are taken by prominent lineage headmen. Women's power remains indirect, but is potent nevertheless. This point is well illustrated by one commercial traveller who commented that several schemes with which he has dealings have not been commercially viable because males have neglected to consult the women in the scheme on important matters, and their resultant lack of support has proven critical.

Heightened awareness of female competence, which is becoming a general trend in Swaziland, is reflected in the respect the members have for some women in authority. For example, a Swazi woman who represents a Canadian lending agency and is always treated with pronounced respect on her visits. She is held in high regard because she is seen as 'honest'. This is a quality many members state is lacking in many male bureaucrats. One man argued that '... women do not use their power to blackmail us'.

Many people believe that women are naturally good commercial farmers. For example, De Vletter (1986) reasoned that the role of women in the irrigation schemes he surveyed is important since '... they seem to have conditioning which makes them very steady'. While this seems to me to be the case, some women resent the extra workload this steadiness entails. One woman, a prominent Real Farmer, told me quite directly that she resents having to do so much domestic work after a day of hard work at her garden.

Such is the importance attributed to female membership and labour that the Real Farmers express strong support for equal political rights for women.[4] A Real Farmer stated:

> This was the tradition long ago, now they are treated as equal to men. Now they are treated equally as here in the scheme

> [Ntimane], there is no male or female, just members everyone. [Concerning a female chairman] If the members voted for her, there would be no objection she would have to be chairman. [When it was suggested that there would be a bias against her] That idea delays our progress. If they say that 'we won't choose a woman to be the chairperson', but if she has the qualities of being the head, obviously they won't progress.

Another prominent Real Farmer reacted vigorously when I suggested that perhaps women would not make good leaders. 'You are unfortunate because you obviously didn't see them carrying the pipes'.[5] When I denied this he commented 'The women are powerful and they are hardworking. They are also equal in terms of their leadership ability'.

This view of women is confirmed by another prominent Real Farmer who stated '... now the time has come where everyone is equal. As in the scheme [Ntimane], there is no female or male, everyone is equal, they are all equal'.

When I questioned this Real Farmer specifically about the Swazi law of female minority upon marriage he stated '... it is a bad thing this law. They [the married couple] should both be working to support the family; I cannot side with this law'.

Real Farmers are very conscious of the importance of women at Ntimane. This awareness is enhanced by their appreciation of the value of labour, and so a relationship between political rights and labour is firmly established. This awareness flows on into a general cognizance of the productivity of labour. The idea of productivity becomes the yardstick for further determining management differences between Ntimane and the homestead. This difference in management was partly illustrated in my earlier example of the female manager at KaShongwe, who uses kin based labour at her homestead, and her own intensive labour at Ntimane. By recognizing that commercially productive labour incorporates some rights to the resultant produce, the Real Farmers have been forced to take remedial action against too much local labour becoming available, which would otherwise reduce their ability to under use their homestead land. This would happen if they employed too much labour at Ntimane, because then they would be seen to have the time to cultivate their homestead land.

The Real Farmers are having to resist demands to employ their kin at Ntimane. In recent years it has become much harder for young people to obtain work in the towns, and there is growing pressure on rural farmers to provide employment. This strategy has worked in that, as yet, there is no expectation that Swazi youths should be employed at Ntimane by their kin. This does place much stress on the ability of the individual Real Farmer to carry out all the labour needed on the plot. By restricting commercial labour management to Ntimane, and using only their immediate family and maybe some *Shangaan* contract labourers, the members can avoid having to accommodate the demands of kin for work. By employing kin at the homestead on essentially non-commercial labour, and paying money either directly to the headman of that family, or into education, there can be no suggestion of employing kin for personal commercial gain. A semblance of balance in terms of status and authority can thus be maintained within the structure of the kin group.

In dichotomizing between the use of labour in the two places, the Real Farmers remain consistent in their determination to gain proprietorial control over land. By their activities at the homestead, they manage to assuage the fears of kin and neighbours that they are gaining some form of advantage, while continuing to under use land through 'inefficient' use of labour in that place, a situation only economically sustained through their intensive use of personal labour at Ntimane. Such is the pressure to labour at Ntimane in person that it limits their ability to expand, and thus reduces competition among members.

A good example of this process is one prominent Real Farmer and his wife, both salaried government employees and wealthy by local standards, who have two gardens. If they cannot use their sons to carry out the work at the gardens, they do it themselves after work hours, even though they are wealthy enough to afford several labourers. By labouring themselves they are able to demonstrate that they are not seeking unfair advantage over others. By not employing labour from outside their family, they avoid entering into relationships by which they would be seen to dominate, or even exploit, those about them.

In managing labour, the Real Farmers must reconcile popular perceptions of traditional labour roles so that they are seen to act in an appropriate manner, both at the homestead and at Ntimane. In the particular case of the division of labour by gender, it is possible that role stereotyping has been previously overstated by Kuper (1947) and Marwick (1966). On their visits to Swazi Nation Land, both stayed in royal homesteads where, as one informant stated '... they have enough people so that everyone can do the proper thing'. For example, the Chief's homestead still operates along 'traditional' lines, and the members of that homestead are expected to live a very conservative lifestyle.[6] When I visited the Chief at his homestead I noticed that, with the exception of the Chief's hut, all the buildings are conical, thatched, and of mud and stone construction. Each wife has her own hut and these huts are surrounded with reed walls that form house compounds. The homestead conforms closely to the model of the traditional homestead which I described in Chapter One.

I concluded that, contrary to Kuper's and Marwick's interpretation of role stereotyping, Swazi commoners are very pragmatic about the demands of everyday life which makes some labour distinctions impossible. For example, some old men told me that the cattle taboo against women lapsed when men first went to the mines (in the early 1900s), and this lapse is reinforced by children going to school. I saw women handling cattle, milking, drenching, and harnessing them. Girls routinely herd cattle, and women use oxen to plough. Role choices have perhaps never been fully available to commoners, and certainly the emphasis the Real Farmers place on practicality has strong popular support in their organization of economic priorities.[7]

The Real Farmers manipulate labour through their ability to relate traditional labour roles and obligations (however defined), to those whom they believe to be currently appropriate. Even more importantly, the Real Farmers are able to convince their compatriots that what they do does not threaten the social stability of Ngwempisi.

Market transactions

Market transactions have occurred in Swaziland since the early slave trade with the Boers, and then when Swazi labourers first began to work on the mines in South Africa. In recent years, with the expansion of Ntimane, the Real Farmers have become involved with big markets which they both developed and dominate in the local area. With marketing behaviour, as has been shown with labour management, the Real Farmers act differently at Ntimane than they do at their homesteads.

Marketing behaviour at both Ntimane and the homesteads can be grouped in three categories. First, commercial transactions that occur with outsiders;

second, semi-commercial transactions which people refer to as commercial transactions, but which incorporate varying degrees of social obligation and redistribution; and finally, ritual exchanges which are limited to the homestead, but have come to involve a significant, though indirect, financial element.

Commercial transactions

At Ntimane, all customers in commercial and semi-commercial transactions are known as *emacustomer* (customers). Commercial transactions are made with customers called 'middlemen', who arrive from other parts of Swaziland and from South Africa to buy staple vegetables (cabbages and potatoes) by the *bakkie* (small truck) load or, on some occasions, by the truck load. These men then sell vegetables to market women at the market centres, who in turn package the product and sell directly to the consumer.

There is also a commercial market for non-staple, exotic vegetables, such as eggplant and chilli, which farmers describe as being for 'personal consumption'. These items fetch a high price, both from the middlemen and from local customers who may be seeking luxury items for special occasions.

In commercial transactions with the middlemen, the farmer receives approximately one third of the final price. For example in 1985/86, a cabbage which sold to a middleman for thirty-five cents would then be sold to a market woman for approximately sixty cents, before finally being sold to the consumer for one lilangeni.

A subsidiary form of commercial transaction occurs when a producer sells directly to the hardheaded market women. Since this requires a motor vehicle it is an irregular occurrence. At Ntimane, during my fieldwork only one farmer qualified in this category and he tended to restrict this form of marketing to coincide with his visits to town, explaining that he covered the costs of travel in this manner. The commercial transactions with middlemen are predominantly formal transactions, though some middlemen patronise particular farmers. The direct sales to market women are highly formal, as these women are regarded as notoriously hard bargainers, and the vegetables are always offered to the highest bidder on the open market.

This form of commercial transaction, the 'small farmer channel', has recently been observed in Swaziland by Sandee (1985, p.22). Sandee (1985) further identified the 'Manzini centre channel' as being contractual arrangements between producers and retail outlets (1985, p.23). At Ntimane only one prominent Real Farmer fits into this category, because of his long established supply contracts with the hospital and the school in Mliba.

Commercial transactions are coveted by the farmers, but involve greater risk, with the possibility of non-sale or even over supply. Commercial transactions involve larger regular quantities and entail a major commitment to mono-cropping, a practice generally avoided.

Semi-commercial transactions with local people are the dominant form of marketing at Ntimane. Though these transactions are called 'commercial' sales by the farmers, in my opinion they are more properly identified as 'semi-commercial', for social obligations often cause farmers to give away far more vegetables than they might in a purely commercial transaction. These transactions, as with the limitation on the use of labour, constitute both a restriction on commercial expansion and a form of security for the farmers at Ntimane.

Semi-commercial transactions

When Ntimane first began, vegetables such as cabbages, carrots and potatoes were not part of the local diet. The availability of these vegetables soon resulted in a situation where, of the 107 homesteads in the immediate vicinity, the seventy-three non-member homesteads have become reliant upon Ntimane for a major part of their diet.

These direct sales to local buyers are the most significant sales for Ntimane farmers, and are the category about which data are scarce, though Sandee (1985) speculates that the irregularity of vegetable supply to central markets may be partly due to the influence of these local sales (p.23). The local sales are crucial to the Real Farmers because such sales are reliable and constant. However, local sales pose a real dilemma for the Ntimane farmers, for, although such sales ensure their success, a preponderance of social ties and social obligation forces them to sell at consistently low prices. Raising local prices would inevitably undermine that market, and even encourage other local people to commence growing, thus forcing existing farmers to become more dependent on external commercial markets. Under producing for the local market would ultimately have the same effect. Therefore the farmers must maintain equilibrium by ensuring an adequate supply of cheap vegetables for local sales, while producing enough surplus, using their limited labour, to make a good commercial sale with a middleman or a market woman. Price rises would therefore be gradual, and local customers would not be encouraged to produce their own vegetables.

Though local sales are vital, most farmers aspire to move away from this state of dependence upon their neighbours. However, previous alternative attempts at bulk marketing in a town centre have failed, after complaints from consumers of poor quality control. This led to general recognition among members that quality control, as determined on the open market, is a strength, since the failings of individuals could not reflect upon the entire organization. This was a constantly reiterated theme among the Real Farmers. Despite this, there is pressure from some younger educated people, and from the Rural Development Area officials, to enter into some form of more centralized marketing operation. The older, more pragmatic members are keen to acquire a truck for cooperative cartage, but on the same basis as tractor hire, without any attempt at communal marketing. They would prefer a situation where they could hire the truck in conjunction with some others, cart their produce to town and then market the produce separately.

The dependability of local sales has many limitations for the farmers, as local customers are quick to press for the best vegetables at the lowest possible prices. Up to one quarter of any crop may be given away, and the customers exert considerable pressure on farmers to give them, without payment, vegetables of inferior quality which may have been previously rejected by middlemen. Local customers, being non-members, often stress that they are less fortunate than the farmers and so gain an immediate advantage, as well as setting a precedent for the future sales relationship. In this situation, the customers' negotiating positions are enhanced by their local knowledge of the farmer's personal circumstances. In a ironic reversal of roles, the customers are able to use Real Farmer values relating to commodity production to demonstrate their own relative inequality, and the full extent of their needs compared to the Real Farmers.

Because farmers have no real negotiation points upon which to build a price with local customers, this relationship at times verges on the exploitative, since local customers force limited redistribution on the Real Farmers. However, it may be that this relationship provides a localized redistribution mechanism, through which the people in increasingly economically marginal homesteads are able to gain access to edible commodities. This is an argument strengthened by reference to a traditional relationship within the hierarchical Swazi society, which allowed individuals to *cela* [*kucela*, lit: to beg] from others without stigma. Kuper (1963) describes this form of customary behaviour in the following terms:

> ... generosity [is] the hallmark of achievement and the primary virtue of *ubuntfu* (humanity) A donor must always belittle his gift, while the recipient must exaggerate its importance ... Begging has a connotation different from that expressed in the European milieu. Among conservatives it carries no shame. To beg is a sign of deference and to give is a token of superiority, enhancing status (1963, p.49).

Because of this strong emphasis on the right to beg, the relationship between the farmer and the local customer involves a certain degree of localized status attribution. As a result, the more a farmer produces, the more pressure will inevitably be exerted on him or her to redistribute vegetables, through the right of others to beg. This process is further exacerbated if the farmer already has a high status.

The pressure to share resources does not always take the form of negotiating a cheap price for vegetables, but may involve other commodities. For example, two married and moderately affluent Real Farmers who own a motor vehicle are constantly under pressure to spread their resources very widely among kin. Not only are they expected to provide cheap vegetables, they are given responsibility for the education of the sons and daughters of siblings less wealthy than themselves. Their motor vehicle is in constant demand for many reasons, few of which draw an immediate monetary return.[8] The over riding assumption among the general community is that these Real Farmers are able to, and therefore should, bear the cost of these constant demands. It is hard for these Real Farmers to resist this pressure from their neighbours. They do receive social esteem for their actions, which is some recompense, but an even more powerful stimulus to carry out these demands is the fear that refusal might lead to witchcraft. For example, this family has already been victim to alleged witchcraft, when the wife was bewitched by a male relative. The issue was solved, but only after a debilitating illness and much fear and resentment. I was told that the alleged offender had been jealous of his wealthy relatives and had sought to kill the woman.

Fear of witchcraft is not the only means by which a redistribution of vegetables and resources occurs. Other members at Ntimane also exercise pressure on fellow members to accommodate the needs of the local customers, since satisfied customers are crucial to the organization. At the same time, the members of Ntimane, as Real Farmers, are eager to maintain good relations with the general community, and so cheap vegetables are a means by which they can demonstrate that they are not a threat to social order.

These expectations to accommodate customers are generally recognized. On one occasion I witnessed a young man, working his parents' garden, who was approached by a *sangoma* (soothsayer) wishing to buy sweet potato. She stated a kin relationship with his father, and then closely scrutinized the potatoes as he put them in her bag, often rejecting a potato and taking another instead. When the required number of potatoes had been reached and her money was due, she returned to a recital of her relationship with the young man's father, and continued to help herself to potatoes that she described as 'small' and 'worthless'. When she finally left she was carrying twice the quantity of potatoes she had initially asked to buy, but had still only paid the original price. Admitting that he did not even know if the woman was related to his father, the young man gave me two reasons for his reluctance to push a harder sale. He told me that his father would be angry if he offended a relative, but more significantly, since the woman is a soothsayer with ritual power, he believed that she should be treated with caution.

Not all Real Farmers are subject to such manipulation. The most prominent Real Farmer at Ntimane is renowned, and admired, for his ability to handle such requests to his own advantage. He refuses overly demanding customers with impunity, and receives general support in his actions from other members. He seems indifferent to the possibility of witchcraft, and is treated with respect by customers. His ability to gain support from fellow members for his actions rests partly on his eloquence, and partly on his prominence in local politics. Owing to his influence as a member of the local Land Management Committee, and over the Chief, any accusations against him would be directed through the Chief, and would involve substantial risk on the part of the accuser.

Other mechanisms exist in Ntimane which exert pressures on members to over produce and redistribute produce to local customers as a means of sustaining the organization. For example, further stability is maintained by reducing competition, such that members do not compete for customers, thus ensuring a high degree of internal cohesion within the organization.

This levelling mechanism makes cooperative work possible without introducing unlimited competition. For example, while marketing is individualized, farmers do not actively solicit custom. Rather, inquiries are made generally and publicly, and customers are directed to the farmer with that item in stock. This is usual with transactions involving middlemen. Customers may approach a particular favoured farmer, but farmers do not actively seek customers. Farmers with specific vegetables in stock prefer not to reply to a general inquiry, but would rather allow others to refer the customer on to them. In this way the Real Farmers reassure one another that individuals are not seeking an advantage in the common marketplace.

Further limitations on individual initiative are evident. Single lot mono-cropping of staple vegetables, a situation where one crop predominates and all plants are the same age, is a device which allows farmers to sell large orders to middlemen, but it also forces farmers to sell, or give, the inevitable surplus to local customers. Even then, a large percentage of the rotting crop may be ploughed back into the garden as compost. This practice reduces the chances of any one farmer constantly having a particular crop on hand, and so dominating a particular part of the market, thus allowing an opportunity for other farmers to make large sales. I came to the conclusion that any large sale to a middleman was put over and received as a matter of good fortune rather than strategy.

An even more subtle self imposed limitation at Ntimane is evident in the pressure exerted by the members on one another to use all their land. Under the constitution, members who under use land are subject to sanction and possible eviction. There is considerable pressure to have the entire plot in use most of the time. This has a dual effect, by first ensuring that a disposable surplus is always evident and can be given away or sold cheaply, and second, ensuring that members are fully occupied at Ntimane and cannot be accused of having the time to better utilize their homestead fields.

Competition for the local market is further reduced by the over supply of vegetables. This also forces farmers to seek out external markets which do not involve direct sales at Ntimane, where the local market cannot significantly increase, or where a share of that market cannot be gained without increasing competition among the members. To this time farmers have not tried to avoid the pressures of local consumers by cultivating non-edible crops for external sale. I concluded that this was because non-edible crops require larger acreages than are available at Ntimane, but also non-edible crops service a market over which the Real Farmers have no control. Because of these factors, the Real Farmers accept the advantages and disadvantages of their relationship with their customers.

Ritual transactions

I have previously argued that Real Farmers treat the homesteads and Ntimane differently when they deal with ritual and market transactions. Ritual transactions only happen in the homestead arena and involve the giving of cattle. However, by restricting ritual commodities (cattle) to the homestead arena, the Real Farmers are using their homestead land to support the cattle they are now using to obtain loans for ventures at Ntimane.

They are able to do this because the exchange of marriage cattle remains fundamental in Swazi social life, constituting a theoretically non-commercial ideal in contrast to the commercial nature of most other transactions. Implicit in the giving of marriage cattle, are rights in uxorem and rights in genetricem, rights of filiation which are achieved over time, and bring children to the patrilineage through the transfer of cattle. The importance of marriage cattle is as current today as it was when Kuper (1959) described it in these terms:

> The indisputable, but by no means sole, evidence of legal marriage is the acceptance by the bride's people of a number of cattle known as *emabeka* under the custom of *ukulobola* ... A man may *lobola* a woman long after she has borne him children, and thereby establish them as his legitimate offspring, but - and in this the Swazi differ markedly from neighbouring Bantu tribes - he may simply *lobola* the children, leaving the woman free to marry elsewhere (1959, p.88).

The exchange of marriage cattle remains a serious affair among the Swazi, a fact reaffirmed by the emphasis, in all homesteads, on cattle ownership. Their control over marriage cattle is the means by which older people reinforce the loyalties of the young, or gain an *inhlanti*, (a child)[9] from those who need corporate help in gathering sufficient marriage cattle. The Chief uses the transfer of marriage cattle as a means of measuring the strength of linkages between patrilineages. His understanding of the degree to which an exchange of marriage cattle has occurred thus provides him with a guide to arbitration in many disputes.

Marriage cattle exchanges are considered to be non-commercial for a variety of reasons. Marriage cattle need never be given in full and can only be given once in their entirety. Therefore, unlike dowry, once marriage cattle are given for one marriage, no cattle can be given for any future remarriage. The primary emphasis in marriage cattle arrangements is on the circulation of cattle and the acquisition of wives and children, rather than the accumulation of an asset. This interest in the circulation of cattle can be seen in the generally poor quality cattle that people own. Cattle are inevitably indifferent and scrawny nguni, with low reproductive rates and poor milk production. The cattle survive on marginal and over grazed land, they have little energy, are docile and are treated for disease only in extreme circumstances. Any attempts to improve stock quality have generally been received without enthusiasm. For example, the High Veldt Ranch, one of the USAID breeding and fattening complexes, offers a variety of services to local farmers wishing to improve their stock. The farmers have remained entirely indifferent to these opportunities. Though the Swazi will state that they 'love' their cows, this takes the form of neglect if measured in European terms. It seems that the contradiction forms the basis for frustrated misunderstandings by officials attempting to prevent over grazing.

The Swazi in fact 'love' what their cows represent, which is the potential for wives and children. In these terms, cattle ownership involves generalized notions of ownership as residing in the lineage headman and inherited by his heir, with obligations to provide marriage cattle when needed, and *tinhlawulo* (cattle given as reparation for an unapproved pregnancy) if a son impregnates an non-approved lover.

> In short, wealth ... in traditional Bantu society represents primarily the actual or potential capacity of the family or lineal kin-group to reproduce itself *[sic]* (Holleman, 1962, p.12).

In ritual terms the animal itself is of fixed economic value regardless of body weight. This set value has also been incorporated into very limited cash marriage cattle transactions where money is nominated as being 'cattle'. This process has previously been commented on among the Pedi (Sansom, 1976, 1981), and among the Shona (Holleman, 1962).

Though conservative Swazi deny the possibility of using money exchanges in the place of cattle, some rich urbanized Swazi resort to this technique. One local youth, in giving marriage cattle on behalf of his uncle, actually gave cash, though two cattle were given as well.[10] On this occasion the marriage cattle were given over in cash, and in their entirety.[11] Cash transactions continue to be generally unpopular, because some headmen are beginning to conduct the affairs as commercial transactions. This is considered humiliating for women, and is also introducing a competitive element into the process of marriage.

In the meantime, the farmers seek an optimum level of vitality in their animals, which facilitates reproduction and milk supply, while allowing easy management by the herdboys, since strong vigorous cattle break the cattle pens and fences, are dangerous to children, and eat too much. Despite these factors the Swazi are aware of the market value of cattle. Butcheries are common and beef is a preferred food. Also, stock theft is usual and sometimes occurs on a huge scale.[12]

The institution of cattle ownership has become a major area of Real Farmer control and influence, for an opportunity has arisen for them to gain access to credit with which to finance ventures on Swazi Nation Land, using cattle as collateral. Because cattle are concentrated around homesteads, it remains in the interests of the Real Farmers to stress their traditional rights associated with cattle. To illustrate, if a farmer has lent money using his cattle as collateral, and has defaulted on that loan, he can avoid losing the cattle by claiming they are only his under cattle lending arrangements. On the other hand, if he needs a loan and he has cattle that are only his through cattle lending arrangements, he can claim those cattle to be his own for the purpose of getting the loan. All over Swaziland, the result has been a spate of 'loan indiscipline', a euphemism used to explain non-payment of loans to lending agencies. The major lender of money accepting cattle as collateral is the Swaziland Development and Savings Bank. By 1985, the Bank was increasingly unable to service loans because it was unable to seize the collateral (Times of Swaziland, 27/12/85). Such was the extent of loan indiscipline that the bank ceased lending to individuals, and chose only to lend to corporations such as Ntimane.

The response of the Real Farmers at Ntimane is not to practise loan indiscipline, but rather to service the loan promptly, and add further credibility to their borrowing record. Using such a loan the members of Ntimane bought a trailer to complement their tractor, paying E2,600 immediately out of the general account, and E1,400 through a loan on the cattle of certain prominent Real Farmers, in a subtle display of wealth. This process immediately established Ntimane as a good credit risk, and reinforced the dangers of lending to individuals on the basis of cattle collateral. The owners of the cattle received the acclaim of their fellow members in an unusual situation which allowed a display of wealth without disapproval, adding to their high public profiles and ultimately enhancing their influence. I drew the conclusion that, by servicing their loan, the Real Farmers at Ntimane consider that the long term gain of good credit status is preferable to the short term benefit of loan indiscipline.

By using cattle as collateral, farmers also avoid the necessity of having to use their crops for that purpose, a prospect which they consider undesirable. Several farmers told me that such forms of credit are inflexible and, if crops fail, would leave them hopelessly in debt. By using cattle, the farmers are using a form of collateral that is renewable through kinship relations, without incurring immediate personal financial strain. Also, through cattle lending arrangements, most Real Farmers have a comprehensive holding of cattle, and can borrow other cattle against those they currently manage. Therefore, cattle are a highly flexible, non-inflationary, and eminently renewable resource that is independent of market forces, and is renewable through social processes at which the Real Farmers are adept.

By restricting cattle to the homestead domain of operations, and by resisting the Rural Development Area Programme's pressure to reduce cattle numbers on Swazi Nation Land so as to put the land to more 'profitable' forms of use, the farmers have increased their cattle numbers. When no diminution of cattle numbers occurred, the government instituted restrictions, limiting homesteads to fifteen cattle. The result has been an upsurge in cattle lending arrangements in the low veldt, where cattle numbers are lower. By continuing to herd cattle at the homesteads, which are protected by traditional rights to graze cattle, but being willing nevertheless to use them as collateral, the Real Farmers are able to use their traditional right to the general resources of the community to provide a credit resource base for their activities at Ntimane.

Conclusion

In this chapter I have argued that the Real Farmers distinguish between Ntimane labour and homestead labour. Kin based labour is used at the homestead for essentially non-productive work, but is financed by work at Ntimane, where virtually all the labour is supplied by the plot owners themselves. This situation results in plentiful labour at the homestead, with little direct commercial return, and scarce labour at the gardens with a very high return, and provides a justification for not being able to apply that scarce labour to the homestead.

Exchanges have been identified as commercial, semi-commercial and ritualized, incorporating strong elements of an ostensible 'multi-centric' economy (Bohannan, 1959, p.492). Of the three categories, commercial exchanges are actually considered least important, while semi-commercial exchanges force a generally equitable redistribution of vegetables to local consumers. By emphasizing the semi-commercial exchange as fundamental in Ntimane market behaviour, the Real Farmers continue to dominate commercial activities in Ngwempisi, while also contributing to harmony in the local social order. Of crucial importance, however, is the way the Real Farmers are using a highly ritually significant item, cattle, to gain access to credit. By refusing to commercialize cattle husbandry, and by restricting cattle activities to the homestead, the Real Farmers have successfully reinforced traditional rights to graze cattle on common land. In this way they support their commercial enterprises on the privately held land at Ntimane, thus allowing them to sustain their under use of homestead land, in their drive for proprietorial rights.

Notes

1. I have designated labour arrangements as 'formal' and 'informal'. Formal arrangements are those in which labour is hired under a form of contractual arrangement. Informal arrangements are those, usually involving kin, where payment is made in a variety of ways determined by agreement with the worker's headman.

2. Because of the civil war in Mozambique, young refugee men come into the area seeking work. During my time in Ngwempisi there were about forty refugees in the immediate vicinity, most in their late teens and early twenties.
3. LaMavimbela, wife at KaMamba, is a forceful woman and head of the Cattle Dip Committee, and occupies a position of power and influence. The dip is the single place to which all people must come on a regular basis, and it has become the place where messages from the Chief are given.
4. In Swaziland a woman reverts to minority status on marriage.
5. The pipes to which he refers are large heavy twelve inch diameter steel water pipes.
6. The idea of marrying a prominent royal family member is not popular among young women, as royal wives are expected to sit on the floor and dress in a traditional style.
7. Sahlins (1972) argued that the household is subject to a cyclical routine wherein, at various times, the household has differing capacities to supply requisite labour. The royal homesteads usually always have enough labour because chiefs are polygynous.
8. Such are the demands on this man that he has even travelled to Durban, in South Africa, to bring back the body of another man's relative for burial, a round trip of some 1000 kilometres.
9. The aged have a right to call on any member of either patrilineage who has eaten the *kugege* (cattle given to console the mother of the girl), for support in their old age. Such help is usually provided by giving an *inhlanti* (lit; 'fish' or 'to wash') who is a small child provided to help with chores, to a young couple without children, or an old person by an obligated couple, to help with chores.
10. One cow was slaughtered and eaten to celebrate the occasion, whilst the other cow was given to the bride's mother to compensate her for the loss of her daughter. These cattle are not part of the marriage cattle.
11. The cash transaction is called 'marriage cattle' (*lobola*), but is calculated at less than the market value of live cattle. Since marriage cattle is given over time, value is enhanced through reproduction. Cash given is therefore less than live value but must be given in its entirety so that cattle can be bought immediately and commence reproduction.

12. In 1986 large scale theft from the King's herds by the appointed guardians of the cattle, for their own financial gain, resulted in a national scandal (Swazi Observer, 18/10/85; Times of Swaziland, 22/11/85).

5 Kinship, lineage and locality

Introduction

The Rural Development Areas Programme enabled individuals to exercise greater control over their ability to gain status and power. The Real Farmers seized this opportunity, and have increasingly promoted, as legitimate, the acquisition of status and power through individual merit and initiative. This is a radical departure from the traditional past when a person's ability to gain status and power was dependent more on clan and lineage affiliation than on individual initiative.

The first part of this chapter examines the way in which the Real Farmers de-emphasize lineage and locality as the major criteria for gaining rights in land, while at the same time using these factors to maintain stability at Ntimane. The second part specifically examines changing marriage relations, and the way the Real Farmers support the economic role of the conjugal family unit.

At Ntimane, the Real Farmers de-emphasize kinship as a criterion of eligibility for entry to the scheme, but then use kinship as an element of control. They also downgrade their commercial activity at the homestead, and so provide no encouragement to others to imitate them, but then use the wealthier Ntimane subsidized homestead to finance and develop a higher social profile.

The Real Farmers have glossed over differences between themselves and members of the general community. Nucleated families, in turn, support the selective use of labour at Ntimane, and allow the Real Farmers to finance, maintain and reinforce their social and political activity on Swazi Nation Land, and their long term proprietorial ambitions concerning land. Ultimately the rewards for the children of the nuclear families lie in an inheritance of land, thus increasing the long term need to own the land.

Kinship

Kinship in traditional society

Swazi kinship is primarily defined in terms of sociological function, though relationships are fictionalized as 'blood' ties and obligations are sustained through continuity in relationships (Kuper, 1947, p.105-108). Kuper describes this in the following terms:

> The nucleus of each homestead is the biological family, established by marriage, but dependent for its continuity on the concept of 'one blood' ... Although the primary relationship in kinship can be expressed in terms of perpetuation through consanguinity, the blood tie is not an inevitable kinship relationship, nor does the same degree of consanguinity with different kinsmen receive the same social recognition (Kuper, 1947, p.105).

Despite this emphasis on fictive kinship, the notion of a 'blood' relationship is ritually significant. Red ochre, representing the 'blood' of the lineage, is still used in the marriage ceremony, and sorcery is particularly potent if the sorcerer and victim have common 'blood'. Though actual blood, as associated with menstruation and miscarriages, is considered polluting in certain circumstances, '..."blood" *per se* has no special mystic significance; and "blood" ties are ignored and kinship ties are artificially created' (Kuper, 1947, p.108).

According to Kuper, locality became important in determining kinship after the breakdown of the clan based chiefdoms, as fictive kinship ties were created among neighbours. Nevertheless, consanguinity remained important; privilege was inherited, and identity rested in clan affiliation with the complementary practice of clan exogamy.[1]

Clans were graded within the national hierarchy by the relative status of their most prominent lineages, and by the way those lineages were linked to other senior lineages. The Dlamini clan was at the apex of this hierarchy, and the king's lineage was pre-eminent. Within a particular clan, members were graded by their distance from the head of their lineage, and then by reference to the closeness of their lineage to the senior lineage of their clan (Kuper, 1947, p.111).

Within the clan hierarchy, consanguinity was considered important in attaining status positions, though clan mobility did occur. Clans that provided queen mothers subsequently enjoyed increased prestige, as did those clans with a strong historical and ritual identity. A low clan status was difficult to improve but not inevitably so (Kuper, 1947, p.112-113). 'Blood' therefore constrained social behaviour, but was not a determinant of it. Politically, consanguinity was most important in the upper status levels of the prominent lineages, while the analogy of the nation as a kinship group, with the aristocracy as the core lineages, was sustained.

> The clan hierarchy is therefore neither precise nor static. While some clans have risen in rank through diplomacy or loyalty, others have been degraded through conquest or the treachery of their representatives. Grading of clans is no way analogous to the caste system. Clan exogamy is an essential part of the political structure; there is no monopoly of occupations; and a certain degree of social mobility is recognized among the elite (Kuper, 1947, p.113).

The close relationship between clan, locality and lineage diminished after the 'clan period', partly through the number of Dlamini princes appointed to chiefdoms, and possibly because of land partition in 1907. Initially, individuals identified both a clan chief and a political chief. By 1935, most identified their political chief as the primary chief (Kuper, 1947, p.114-115). However, the one important factor which remains current is the relationship between specific localities, and contact with the ancestors. The ancestor:

> ... hovers near its own grave, then, when brought into the homestead, it is at the same time in the company of dead kinsmen whose graves are scattered over the country. The ancestors in fact are everywhere and hear and see everything that is said and done (Kuper, 1947, p.187).

The physical distance of the homestead from the graves of the ancestors does not prevent communication, or prevent the ancestors from watching over the children. It is simply that the ancestors cannot enter homesteads that are too far away. People are not preoccupied by the ancestor cult. They call on them in times of need, but do not feel the need to call on them when times are good (Kuper, 1947, p.187,196; Marwick, 1966, p.221). Individuals do retain a sense of locality, sometimes expressed by reference to the ancestors, but usually in terms of an allegiance to the place of their birth, and have the comfort of a comprehensive array of kin structures which provide for personal security.

Traditionally within chiefdoms, political control was related directly to rights of usufruct over land administered by the local chief. Conformity was premised on the promise of land after giving allegiance. Social and political obligations were implicit on receipt of land, and headmen gave unquestioning support to their chief and to the king, by fully accepting their judicial power. The rights of headmen were an assurance of land, protection from unjust accusations of witchcraft, protection against external depredations, and above all, spiritual and moral support. Influence depended directly on the individual's ability to muster conservative political support. Traditionally, headmen sought contacts and respect for their wisdom, a process facilitated by their clan and lineage status.

The onset of the Rural Development Areas Programme encouraged both men and women to look at other ways of achieving status. People are able to concentrate on commercial activity, and still gain status without having to engage in substantial political activity. Despite this, the Real Farmers tend also to be gregarious, combining commercial activity with conservative social and political behaviour. In achieving this state, the Real Farmers have de-emphasized lineage and locality, stressing instead personal character traits as important factors of eligibility to share common resources. Despite this, they have also reinforced kinship values that facilitate their control over the labour of their fellows.

The question of individual characteristics, as a factor in determining rights of access to land and eligibility for membership of Ntimane, has been previously explored in Chapter Three. What is increasingly evident is that the Real Farmers do not deny kinship, but they have re-defined the boundaries of the community in which consanguinity and fictive kinship have currency. Specifically, the Ntimane Irrigation Scheme has become a community differentiated from the general chiefdom. The difference is in a reversal of procedure, in which differences between Ntimane and the homestead are glossed over at the level of the general community, but emphasized within Ntimane itself.

Ntimane is part of the general community of Ngwempisi, subject to the judicial authority of the chief, and contained within the physical boundaries of the chiefdom. At this level, there is no difference between the status of Ntimane and that of any of the homesteads. Indeed, the Real Farmers propagate that notion. However, in a fundamental way, Ntimane is becoming very different from the homestead. It is taking on the aspect of a community in its own right, with differentiated control over land, a strongly articulated set of community values, an economic support base and a internal, legitimate, judicial authority.

In the process of developing a sense of community focused on Ntimane, the Real Farmers have cut across lineage and locality as factors of eligibility for membership, favouring, instead, personalized qualities of honesty and industry. This emphasis on personal characteristics and lack of emphasis on lineage can be seen, not in the close consanguineal and agnatic relationships between members, but in the number of such relationships between members and non-members, reflecting the closeness of relationships in the general area. For example, among the thirty-five members of Ntimane, thirty-one are closely related to other members, and twenty-three are closely related to non-members. Therefore, most members are not only closely related to each other, but many are also related to non-members. Some of these relationships are very complex. For example, the owner of plot six has a son who owns plot one, a daughter who works with her husband on plot nine, two brothers, one on plot thrity-six and the other on plot thirty-four, a sister on plot forty-two, his brother's son is married to the owner of plot two. He also has three brothers who are not members. A more typical relationship network is that of the owner of plot twenty-nine whose son owns plot thirty-two, this man has a brother who is a non-member, but that brother's daughter is the owner of plot ten.

So closely related are the majority of the people in the Ntimane sector of Ngwempisi that lineage cannot be used as a factor for determining eligibility for membership to Ntimane. My conclusion was subsequently borne out by the emphasis informants placed on individual qualities among members. However, once the selection of members has been made, kinship once again becomes important in maintaining stability within the system. Previously, in Chapter Two, I argued that management strategies at Ntimane rested heavily on consensus among members. This consensus is, in fact, the end result of a process of politicking in which prominent Real Farmers use their sophisticated understandings of kin relationships, which are not confined to Ntimane but are focused on Ntimane, to exert pressure for conformity among members. The outcomes of this process are then ratified at the general meetings.

This political process is best illustrated by reference to the example of KaMotsa, mentioned earlier in Chapter Three. In this case, an aged and increasingly incapable woman, who has a plot at Ntimane, has a son who does not care for her. Recently, a local girl became the son's lover. The old woman now allows the girl to work her plot, and has approval from the members for this. The son is disliked by the members. Members often refer to him, in a contemptuous manner, as a lazy drunkard. In no circumstances would they allow him to assume responsibility for his mother's garden. In another situation, such as at Mzimneni, he would have been her heir under traditional rules of primogeniture. The Ntimane members ignore this rule, and are prepared to take the garden away from the woman, rather than allow the son to join the scheme.

The young woman receives the support of the members for several reasons. First, since the old woman, now her patron, is completely dependent on her, the members can exert strong control over the young woman by influencing the old woman. Second, the young woman is the daughter, by his deceased second wife (KaMamba), of the polygamist owner of plot fifteen (KaMamba), an early and prominent member of Ntimane. Finally, the young woman has a reputation as a hard worker and is quickly clearing the outstanding debt the old woman has incurred.

In the first level of control, the Real Farmers know that they can exercise influence over the young woman's activities through the old woman, who directly benefits from her gardening. If the young woman has a child it will make the Real Farmers' position even stronger because such an event would confirm the grandmother status of the old woman, and inextricably link the economic interests of the old woman and the young woman, regardless of what happened to the son. In this sense the young man is entirely peripheral to the success of the venture, because ultimately any male child would ensure continuity of the lineage and confirm the inevitably of the young woman taking over tenure of the garden.

The second level of control exercised by the members over the young woman is directly through the relationship of father and daughter. As a senior member of Ntimane, Mamba fully appreciates the complexities of organization at Ntimane, and can be relied on to exert pressure on his daughter when required. Finally, the corporate interests of Ntimane are well served by the industry of the young woman. Since Ntimane had carried the old woman's debts for some time, the organization is able to regain some of the outstanding money. More importantly, Ntimane members recognize, in the young woman's ability to work hard, the values they themselves espouse. They are thus able to use the situation to gain a suitable new member, over whom they have control. Clearly, the members of Ntimane are subordinating the corporate interests of the lineage, as represented directly by the old woman's son in this example, to the interests of Ntimane, by recognizing and rewarding the family unit that supports cooperative interests. Should a dispute occur between the young woman and the young man over the use of the garden, the members will support the person who actually works the garden. Despite his familial connection, and Swazi laws of female minority on marriage, the young man will have no claim on either the land, or the produce of that garden. The Real Farmers are thus able to de-emphasize particular traditional rights associated with primogeniture and consanguineal relations, while stressing other aspects of kinship and community which allow for the recruitment of suitable members over whom they can exercise control.

A further area of influence is the manner in which the Real Farmers use the rhetoric of kinship in public statements, but expect members to make individual choices. For example, all farming decisions are individual decisions, and gaining access to plant is based on individual contracts.

However, when required, communal work is handled by work parties, and errant members are subject to fines and criticism for their attitude to the cooperative. This situation reveals how the Real Farmers accentuate individual responsibility, whilst stressing Ntimane as a community in which kinship counts.

The importance of fictive kinship at Ntimane is illustrated by this comment made to me by a prominent Real Farmer '... most people are related and this gives them a good attitude to Ntimane'. A 'good attitude' is an appreciation of collective values, honesty and industry. Considerable skill is needed to sustain the balance between accentuated individuality, community, and kin relations, as illustrated by the following outline of the problem by the leading Real Farmer at Ntimane:

> For people to work jointly they come together by themselves, even if they don't know[2] each other they agree [on common goals] then they are related. We do it [farm cooperatively] because it is within us. But if you call someone and say 'Hey come here and do this [work]' then it is imposed and they won't want to do it. Even if you are related and I impose something that is not within you [of your own initiative], is not pushing you from behind, obviously there is nothing you are going to do, even if you are related.

The notion of community at Ntimane is further reinforced by a high degree of parochial feeling among people who have spent their lives in the area. For example, Ngwempisi people believe that *tokoloshe* (demons) occupy Mt Hokwane, which is only a few kilometres across the river. When I visited the area I found it to be extensively grazed with many adjacent homesteads. Arguably, such parochialism is an artifact of a heightened sense of locality in the past, but is also an emotion that reinforces the identity and interests of local people. This phenomenon is also reflected in the names chosen for various enterprises. As previously stated, schemes in the area gave names to their enterprises with meanings such as 'trying to do well', or 'to open eyes' or 'those seeing ahead'. Even the private farmer, Vilani, calls his garden '*Masitapele*', which means 'you can get what you need', and which indicates to the buyer that he has much produce to sell. By using such names the schemes, and individual farmers, distinguish themselves from the others around them.

Kinship still presents some obstacles to unrestricted individual initiative, and can thus prove a burden. It is burdensome in that clear cases of grievance can be left unredressed because of general concerns about not affronting the kin of the offender, or because of the traditional status of that person, or his lineage, in the general community. Because of these considerations, Real Farmers confessed to me that they had, at times in the past, been unsure of how to act in these circumstances, and retrospectively felt they had left the situation unresolved for too long.

For example, one garden had been owned by a local man who is also a local bureaucrat, has kinsmen at the gardens, and has close marriage connections with the chief. He neglected his garden and was considered by the members to be an apathetic farmer. Though the members wished to expel him, they felt they could not do so without risking the annoyance of the chief and the man's kinsmen at Ntimane. The problem was further complicated by the members' concern to avoid alienating the local Rural Development Areas Programme office where the man is employed. Eventually, to the relief of all, the man decided he had too many responsibilities and gave up his membership of Ntimane.

I concluded that the benefits of kin relations at Ntimane lie in the fact that kinship forms a buffer against hegemony and individual ambition, while providing networks of influence and control that the Real Farmers can use. Social relationships act as a constraint on authoritarian behaviour on the part of Real Farmers or Committee members. Fictive kinship at Ntimane encourages the members to consider themselves a community within Ngwempisi, entry to which requires individuals to demonstrate a sense of community, honesty and hardwork.

Kinship and the conjugal homestead

The homesteads of the Real Farmers, male and female, reflect an emphasis on conjugal families within monogamous marriage relationships. This directly serves Real Farmers' interests in gaining proprietorial control over homestead land by reshaping aspects of the house property complex to focus loyalty on the parents, rather than on the chief. Specifically, in a polygynous marriage relationship the economic interests of the mother and her sons coincide. Politically, the father provides his sons with the crucial linkage to his chief, and therefore the sons come to focus loyalty on the chief, using their father's status to launch themselves into the affairs of the chiefdom.

The most striking aspect of monogamy in the conjugal homesteads is that a commitment to monogamy is physically represented. In an explicit manifestation of intent by the conjugal couple, the very configuration of the homestead is a statement of their marriage ideals. The traditional homestead was laid out in a manner that indicated the often unfulfilled ambition of the headman to have more than one wife. The modern homestead is laid out in a manner that is indisputably a statement of monogamous intent.

The traditional homestead The focal point in the polygynous homestead was the right of the woman, through her traditional rights of wifehood, to have the means to support herself and her children, while the headman retained the right to control the resources of the homestead through his mother (Kuper, 1950, p.92).

The traditional homestead [3] layout, as described by Kuper (1947, p. 38-40; 1963, p.19-22) and Marwick (1966, p.11-31), was stereotyped and standardized. This traditional homestead is illustrated in Figure 5.1. The cattle pen was the focus of the homestead, and the point about which it was aligned. Built facing the sun, the cattle pen was central to the rituals of marriage, death and negotiation. It was the symbol of family, and access to it was controlled by men and older women. Grain was stored in the cattle pen in large underground pits [4] which limited women's access to grain. The cattle pen was reserved for cattle, and all other livestock was penned separately. The construction of the cattle pen, upright sticks buried deep in the ground, was also a traditional form of fortification.

Immediately behind the cattle pen was the *indlunkulu* or *kagogo* (the 'great hut' or 'hut of the grandmother'). This hut was associated specifically with the headman's ancestors and was a symbol of the patrilineage. Without it the homestead did not exist. This hut was presided over by the grandmother who was the headman's mother, or in the event of her death, his first wife, or a wife of the same clan as his mother. The grandmother presided over a realm exclusive to all daughters-in-law, or other co-wives. The grandmother represented her son's patrilineage and her own 'house', for it was her 'house' which became the homestead. The grandmother and the great hut were of special significance in the homestead, since she was grandmother to all the

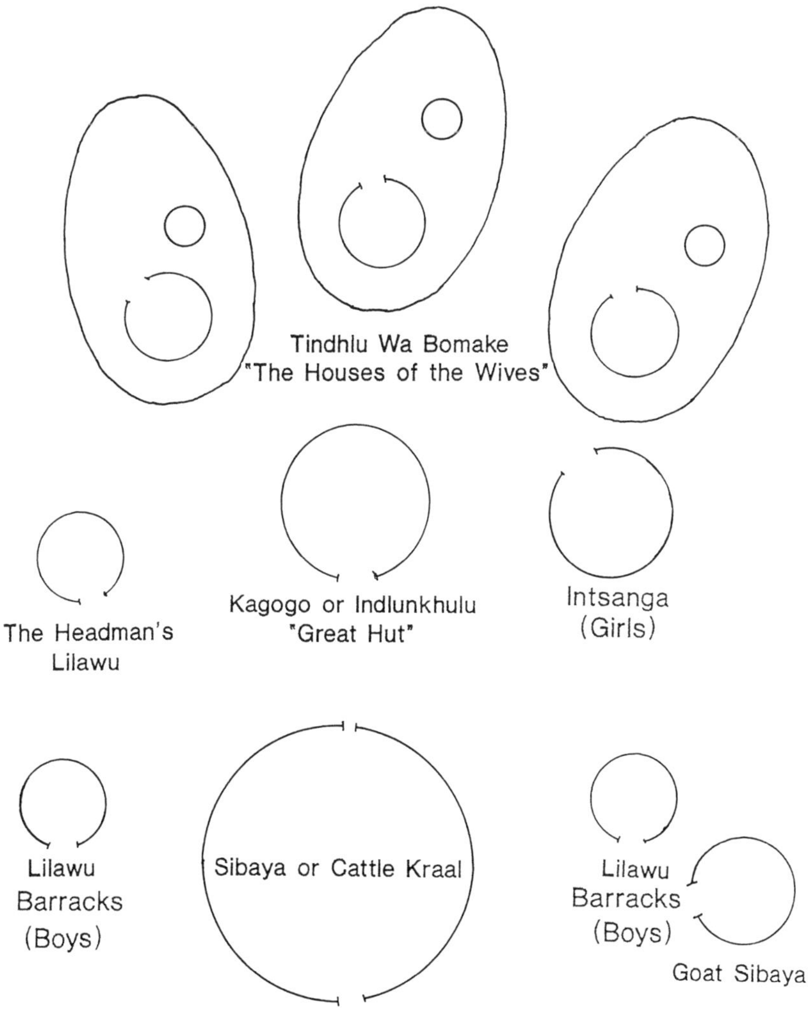

Figure 5.1 A "Traditional" Swazi Homestead
Idealised and Abbreviated Representation

children of her son's wives, and her hut was the one place to which all could come.

Behind the great hut were the houses of the wives. These houses were grouped such that those of the senior wives were closest to the great hut. Each house had its own huts, yards and fields. Married sons built close to their own mother, who would supervise their wives. This ensured that the patrilineage was reinforced among women by a woman whose interests represented the interests of her son. Rigid demarcation between the houses of the wives reduced conflict between co-wives and between wives of the sons.[5]

The headman's hut, is called a *lilawu* (barracks hut), and was situated overlooking the cattle pen. This hut occupied a metaphorical high ground within the homestead, situated to the right of the cattle pen and overlooking the cattle, the compound, and the common ground of the great hut. The headman was fed there by his wives, who shared that task equally, though he would visit his wives in their own compounds at other times. Like the great hut, the headman's hut was a permanent structure.

The huts of children were in various configurations. The *tintsanga* (girls' huts) were the huts for adolescent girls. These huts were situated near those of their mothers, and it was here that they received their lovers. Young boys and young men lived in barracks huts. These huts were situated either side of the cattle pen so that the young men could provide protection for the homestead and for the cattle. These huts were an integral part of warrior socialization, but since the cessation of conflict in the late eighteenth century the have lost their strategic importance.

The traditional homestead was the physical representation of the control of the polygynous family by one male, through the agency of a woman; his 'mother'. The grandmother represented his ancestors and her own immediate familial interests. Her position was strengthened by her occupation of the great hut, which was the common ground for all the children and her grandchildren. The grandmother had the most in common with all members of the entire homestead, with the exception of the headman. However, she was also grandmother to the children of her other sons in other homesteads, and so provided many linkages to other kin. This power was evidenced by the number of children she would bring in from other places. The grandmother also reminded her son of his obligations to his kin as head of the extended family. The grandmother's position was one of great power, attained through her son, personified by the queen regent, and aspired to by most young women.

The contemporary homestead Contemporary homesteads have diversified rapidly from the traditional model described above. While these changes are not easily standardized, I did identify three types of homesteads. The first type resembles the traditional homestead. These homesteads use round huts, thatched roofs and delineated living areas, laid out in a relatively traditional manner around the great hut and the cattle pen. The second style of homestead is the modern homestead, consisting of one main house with many internally connected rooms. This type of homestead does not have a great hut. The intermediate homestead, however, is the most often built and frequently has a great hut and a variety of hut forms. Both the modern and intermediate homesteads rely heavily on concrete block construction and usually have rectangular rather than round huts. This type of homestead is shown in Figure 5.2.

Unlike traditional homesteads, the modern and intermediate homesteads show much evidence of the influence of western material culture. Steel frames and roofs are used, the huts are rectangular to accommodate tables, chairs, and beds, and have locking doors and windows because of the incidence of theft. Homesteads in these two categories may also have a variety of more sophisticated agricultural items, ploughs and metal grain storage tanks.

The kitchens of modern and intermediate homesteads are quite distinct from those of the traditional models. All modern, and most intermediate homesteads, have a large and expensive South African made slow combustion stove.[6] These kitchens are large and display an array of sparkling cooking utensils. Aluminium cooking utensils are common, desired, and coveted. Plastic basins and buckets are sought after, and plates, cups, jugs, and cutlery are prized and exhibited. The food eaten reflects this basic change; baked food is becoming common, maize porridge is losing its popularity, and increasingly, prestige foods are those that can only be created in these new style kitchens. For example, cakes, baked maize bread and scones, which are most popular, cannot be made in a traditional homestead. Women appear to be under pressure from their peers to have modern kitchens, because so much successful socializing is carried out in, or around, such kitchens.

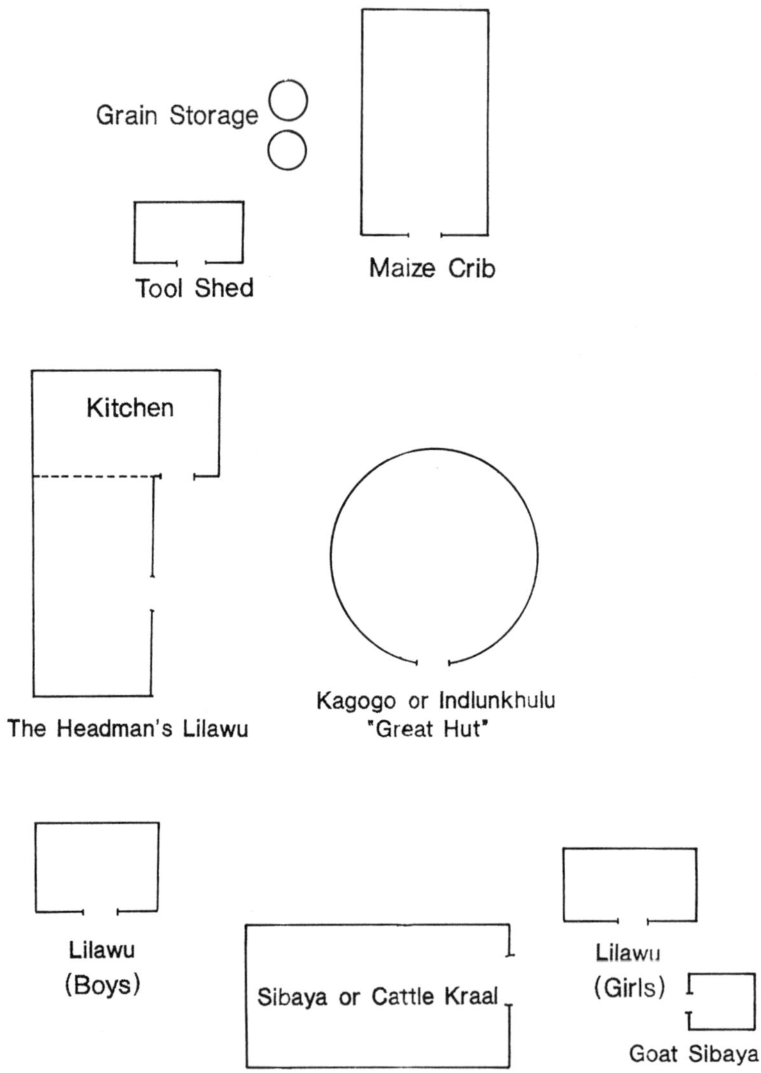

Figure 5.2 An Intermediate Homestead
Idealised and Abbreviated Representation

The change in the configuration of the homesteads is enthusiastically endorsed by many women. This is most obvious when the kitchen becomes the focus of the entire homestead. Previously, in traditional homesteads, the cooking area was enclosed within the house compound of the woman, or if she was an only wife, she might cook near the great hut. If guests arrived, the great hut area was the common meeting area, but in the intermediate homesteads, visitors will sit either in the warm kitchen or just outside it. This redirects the focus of social activity in the homestead back onto the woman, and reinforces her changing status in the homestead and in the marriage relationship.

Many men work in town and the women manage the homestead, dealing with all enquiries and business arrangements, usually at the kitchen door. This brings the business of the homestead directly to the woman's 'house'. What had previously been a private domain has now become public, and the woman who presides over this focal point of the homestead is central to its affairs.

The significance of a commitment to monogamy is immediately evident in the large financial investment made, by the homestead, in the kitchen of one woman. No ordinary polygynist will invest in a concrete block kitchen and a large expensive slow combustion stove for one wife while there is any chance of him acquiring other wives. All modern, and most intermediate, homestead headmen that I knew in Ngwempisi were committed monogamists.

Since the 1930s christianity, taxation and urban drift have discouraged polygyny (Marwick, 1966, p.40).[7] At the same time increased emphasis on consumerism has made polygyny significantly more expensive. Even fifty years ago, according to Marwick (1966), polygynists were finding the practice financially onerous.

> A polygamist can no longer comfortably afford to keep many wives unless he is particularly wealthy, even if he has the initial outlay to pay *lobolo* [8] for them. One man informed me that the wants of women were increasing to an inordinate extent. Instead of being satisfied with a *sidwaba* (skin skirt) once in every four or five years, his wife now demanded less durable

> clothes of a European pattern at short and regular intervals, and no man without means could keep pace with their demands (Marwick, 1966, p.40).

The emphasis by women on expensive kitchens in modern and intermediate homesteads is part of a general trend to female domination of the homestead purse. While it has always been accepted that women dominated the homesteads, these women had been the grandmothers who acted in the interests of their sons, while the wives of the headmen acted in the interests of their own houses. In the monogamous homesteads it is the wives who dominate the homesteads, and decide how the resources should be allocated. The wives of monogamous headmen have the advantage of values that now support monogamy and the right of women to give priority to the needs of their children.

This right, and these values, were illustrated by my previous reference to Real Farmer condemnation of the Swazi law of minority for women on marriage. The Real Farmers, men and women, strongly support the prerogative of women to demand the resources necessary from their husbands to support their children. I was told by one man that he disliked another man who drank heavily, and whose wife worked a garden at Ntimane. He said:

> I could not support this man if he wanted to work on his wife's garden because he would want to take the money. She works hard for her children and he would want the money to drink. I could not support him in this.

Male Real Farmers support this trend to female management at the homesteads, because women readily adopt Real Farmer business practices and ethics. The Real Farmers are thus able to reinforce the traditional direct interests of women in their own houses by promoting monogamy. This emphasis on greater female autonomy is also part of a more general trend in the community. Thus the Real Farmers are not supporting initiatives that are radical, but are astutely recognizing inevitable future realities.

Changes in domestic behaviour, and resultant changes in the status of women, have caused a backlash among some males, and reflect an on going issue with long historical antecedents. There has always been a concern that young women will be lost to the city and will suffer from any loss of social control. For example, Marwick (1966) describes the movement of women into the cities, in the 1930s, in these terms:

> With the weakening of family and tribal sanctions girls and women have drifted to urban areas in the territory and the Union of South Africa, there to become prostitutes or concubines. These women would formerly have found husbands and have been absorbed into the social and economic system as principal characters in a household unit (Marwick, 1966, p.39-40).

Though Marwick's assumptions about women's activities in urban areas are not currently representative of male views of women in Swaziland, many men express concern over the relative freedom women enjoy in urban areas. Many men feel particularly threatened by the great number of young urban working women who are having illegitimate children. Because they are financially independent, and are sometimes home owners under individual title, women are able to avoid marriage by taking many lovers, while retaining their majority status (Armstrong and Nhlapo, 1985, p.26-27).

Many young men expressed to me their uncertainty about the control women exercise over their reproductive capacity. They feel that women are able to draw on male resources without necessarily bearing children in return. They are not confident of the paternity of the children because so many women enter into multiple relationships. Their uncertainty is exacerbated by a generally held belief that the incidence of desertion and divorce of men by women increases as women move to the urban areas. Ironically, there is also a general unwillingness on the part of young men to acknowledge that the traditional values supporting permissive male behaviour, in a polygynist society, might encourage women to avoid any commitment to one man. Young men condemn young women for having many lovers, yet boast of their own prowess and infidelity.

Public reactions against changes in the status of women are common and can take the form of negative media images and overt hostility, particularly from young men, or are reflected in certain government regulations. Many men still try to exert the traditional authority of male kinsmen to question the behaviour of women. While older men can still legitimately exercise this authority, young men are also trying to challenge young women on this basis, an action that generally attracts public disapproval.

Two separate examples illustrate this point. In the first case I shared a twelve hour bus trip to South Africa with approximately sixty market women. The women passed the time chatting, singing, telling jokes, and gambling. After a few hours they became particularly uninhibited in their behaviour. Suddenly, the only other male on the bus, an old zionist carrying his brass cross, stood and firmly rebuked the women for bad behaviour. Immediately all the women modified their behaviour without protest, although they were evidently bemused by the incident. I was surprised at this response, since market women are notoriously strong willed, yet they accepted without protest the old man's right to adjudicate on their behaviour.

On another occasion, I observed a young woman carrying her baby who, when travelling on a local bus, was confronted by a partially intoxicated, aggressive young male who accused her of being absent from her homestead without the permission of a senior kinsman. He was immediately shouted down by other passengers, and the young woman ignored him. It is feasible that the others reacted against his drunkenness, but from their attitude towards him I concluded that they believed him to be acting inappropriately by meddling in the young woman's affairs.

Many young women complained to me and other acquaintances of harassment by young males, and assaults on women in public are relatively common in urban areas. Accusations of prostitution are also commonplace, though they seem to spring more from the frustration of young males than from reality, particularly when young men readily express a desire for sexual contact with many women. Traditional practice had allowed women to take lovers, and men to conduct love affairs with more than one woman.

While still in her teenage years, a woman usually took an 'approved' lover (approved by her family), who would eventually become her husband. As a result most women have children while still young, and usually well before they are officially married. Men have always been expected to be polygynous, so they sought to have more than one wife. These customary practices have collided with urbanization and economic autonomy for women, and in the cities have resulted in many men experiencing confusion and uncertainty over their traditional roles.

In rural areas, where conservative behaviour still applies, young men are expected to desire women. One young man told me that every man is '... driven to desire many women', that this desire is 'natural' and that there is no reason to inform one woman if he is seeing another. He added that unfortunately this is difficult in rural areas because:

> If you arrange to meet a girl near the river, other people will ask her why she is there and very soon everyone will be aware that something is going on. They will tell her family and your family and everyone will be angry that you are having an affair.

In the city, it is possible for men to conduct such affairs without such scrutiny, but it is also possible for women to avoid kin pressure to marry. The institution of the approved lover legitimates these relationships, especially if the inference is that there will eventually be a marriage. Cohabitation is common in towns, and there has been an upsurge in 'christian' (civil service) weddings, owing perhaps to the law that restricts further marriages after a civil service. Church weddings are particularly popular with women, as they can thus guarantee their monogamous status, both legally and religiously.

Other reactions against women's urban freedom take the form of discriminatory government legislation against them and their actions. The most prominent example in recent times was a debate in parliament over women's rights to wear trousers, which are considered 'untraditional'. As a result, women are banned from wearing trousers in government offices.

Further concerns are expressed over women's access to the contraceptive pill, the use of which has been encouraged by family planning agencies. While males resist using condoms, since contraception is a female issue, the contraceptive pill is seen as a threat to male virility. Accusations of licentious behaviour are often levelled at young women, a situation exacerbated by sections of the media, which focus on the negative aspects of female behaviour rather than on similar male behaviour. Press accounts support the right of males to socialize with other women, or to drink, but condemn women for similar behaviour, a view supported by some readers. For example one reader wrote to the Times of Swaziland in the following terms:

> It is a disgrace to find mothers in the *shebeens* [unlicensed drinking places] after 6.00 pm. When do they cook for their families... surely this results in fights with the husbands? I beg the police to raid drinking places all over the country. Women should be given two hours to drink and men four hours. This

> will give women sufficient time to cook for the family, because a man needs hot porridge (Times of Swaziland, 13/6/85).

Media concerns with negative images of females in situations where they exercise personal autonomy have resulted in bizarre headlines. For example one headline read, 'Woman Runs Amok in Nightclub: Guard Bitten on Private Parts' (Times of Swaziland, 18/12/85). Some young male correspondents are notorious for the chauvinism implicit in their representations of women. A case in point was the story of Bongiwe, a young mother deserted by her public servant boyfriend. She received bitter criticism for finally dumping the baby in the father's 'in tray' in an attempt to draw attention to her plight. The correspondent reported:

> After a few drinks we had to do some counselling to our friend. We told him to take it easy as he was the first to face such a situation. Baby dumping is in *[sic]* the increase these days. Former girlfriends have started ruining affairs and harassing their former men.
>
> We also told him to go to his lawyer for advice as the situation was getting out of hand. The girl was now blackmailing him, sucking a lot of money for no apparent reason (Times of Swaziland, 3/4/86).

The media are quick to support government initiatives that stereotype female roles. Even non-government publications give extensive publicity to the work of the various women's organizations, such as the National Association of Swazi Women. These organizations are usually closely associated with particular members of the royal family, and play a prominent part at important national ceremonies where their members wear the traditional costume. In these circumstances, publicity reinforces the stereotyped view of women's roles promulgated by these organizations.

There is general recognition, however, of the degree of responsibility that women exercise over children. Men, as fathers, are not seen as the primary care giver and women are criticized severely if children are neglected. The Real Farmers recognize that this responsibility is a strong motivation for many women. It is this view of women that the Real Farmers propagate, by recognizing the pivotal role women play in managing domestic affairs. The

Real Farmers react strongly against the suggestion that women might be discriminated against in the workforce. They defend the right of women to manage conjugal funds, and they disagree strongly with the Swazi law conferring minority status on women upon marriage. At times, this attitude is supported by media representations, with some correspondents even arguing that Swazi men:

> ... have no appreciation whatsoever of their womenfolk ... what is funny is that our menfolk cannot hear of a Swazi woman marrying a foreigner [in preference to a Swazi] (Times of Swaziland, 18/10/86).

The male Real Farmers support the positive stereotype of women as breadwinners. They support the increased autonomy of women, because it is not a threat to them. Because the Real Farmers are usually christian monogamists, and have a strong commitment to commercial activities on Swazi Nation Land, female emancipation is a benefit to them. Emancipation is attractive to women, and when they become Real Farmers they exhibit the values articulated by the prominent males. The women's sense of personal autonomy is not threatened by the fact that so often it is prominent males who do the talking, as these same males have demonstrated their commitment to a change in the status of women, by enthusiastically adopting women's work.

Changes in domestic behaviour have also influenced the way young men think about their parents. Aware that the chief may not be able to give them land, young men are looking to their parents for land and opportunity. This is made more possible because of the coincidence of the mother's interest in her son (the son of her house) with that of her monogamist husband. Thus united in their interests, the parents are able to focus on providing opportunity for their children, and this is already happening.

For example, one prominent Real Farmer from Motane disclosed that he is going to accommodate his sons on his 'own land' (homestead land). He told me cattle grazing should cease, and all land should be turned over to cultivation, through individual title. Young men should not give allegiance, but should live in villages formed around their parents' homestead.

> We should make things like villages. In this case I suggest that my sons shouldn't go out and *khonta* (give allegiance) but that

they should stay all about my homestead to reduce overcrowding. Now everybody has a big homestead covering a big space.

Another Real Farmer from Ntimane informed me that there should be a strict division of land between cattle herders and commercial farmers, and he cited the late King Sobhuza II:

> In the case of land running out and the fact that they [the farmers] won't have any land, the late King, Sobhuza II, saw that problem and he said: 'Now we have to make stories in the homestead and get electricity and let our children live above us in the stories'. Now the chief does not have much control over that [the process of allowing sons to live in the homestead] because the King mentioned that since the land is running out we need to economise *[sic]* the land.

As in this example, the reference to the King's opinion provides this Real Farmer with a powerful and legitimate reason for seeking more control over his homestead land. As he suggests, the children must focus their ambitions more on their parents, as the Chief can no longer give the assistance they need. This change of focus has had a noticeable effect on public perceptions of the Chief's role, and also on the behaviour of young men who have not yet given allegiance.

Formerly, young men were highly mobile, moving from one kin homestead to another. They identified strongly with their mothers, and their paternal grandmothers. Now young men are focusing more on the parent, or the parents, who control their natal homestead, and are spending much more of their time there. As well, they tend to invest more labour, or remit more money, to the family enterprise, and many plans for the future involve enterprises situated at their fathers' homesteads. One young man told me that he was going to invest in the necessary equipment so that he could become a commercial rabbit producer. His intention was to build a shed at his parent's homestead and breed rabbits in an intensive battery system. This project would involve a relatively large financial outlay and certainly assumes continued tenure of that piece of land by the family.

Another young man told me that he would build his own home at his father's homestead on some waste land set aside for the purpose (to save arable land), because he is considering starting a transport business with his father's support. This particular young man continually works for his parents. He spends alot of time maintaining the homestead, and has designed a small water supply so that his parents can have constant water at their homestead. Subsequently, his parents bought him a motor bike and he uses it to run errands for them. When he graduated from teacher's college he went to considerable trouble to be transferred into Ngwempisi, so that he is able to spend as much time as possible at home.

Changes in domestic behaviour have also affected the relationship between married couples. It is also significant that the traditional reserve between men and women in marriage is being replaced with fellowship. No longer does a woman necessarily fear that she might become a co-wife, and no longer does the man fear that contact with his wife might involve jealousy or financial demands. As a result, I noticed that married couples, especially Real Farmers, worked together in their gardens. Traditionally, men and women maintained distance from each other, and this imperative is still noticeable in social situations. For example, men and women will still sit apart at a meeting, but it is a common sight to see couples sitting together in their gardens quietly talking, or even walking back to their homesteads together.

Despite the maintenance of some traditional rules governing the interaction of married couples in public, the significant difference is in the degree to which couples agree on common goals. Male Real Farmers recognize the need to share decision making with their spouses, because they are dependent on their labour and initiative. For example, I questioned one male Real Farmer, who had his own garden, about his wife's activities at her garden. I was actually interested in how they bought their inputs, but he replied: 'She [his wife] makes all her own decisions about what she grows at her garden. She is a member and this is her right'. I suspected that he thought I was insinuating that he exercised undue influence over his wife. Attempting to redress this misunderstanding, I sought to clarify my interest by asking him if he and his wife shared the purchase of inputs. He answered: 'We are one family; all the money we make is for our children's education. When my wife makes decisions for her garden she thinks of all these things'.

In the above example, the Real Farmer's defensive reply highlighted the degree to which he defended his wife's right to make certain decisions. At the same time he also provided insights into his perception of conjugal responsibilities, and significantly, the importance they both placed on the education of their children.

The Real Farmers further differentiate themselves from others by their high level of participation in local affairs. Even though a wage earner in the city might maintain a high rural profile, it is done with difficulty because wage earners are frequently absent when crises occur. Too often wage earners are capable only of visiting their homesteads for short periods. The Real Farmers are always ready and always present to respond to crises and social initiatives. Real Farmer involvement in local affairs is always expressed through the homestead. Status positions are all associated with the day to day business of the general community. Meetings are held and social occasions occur in one homestead or another. People are constantly moving, visiting, talking and generally socializing. The ambitious seek the role of host, and certain hosts are popular, for their warm kitchens, the good food, and salubrious surroundings. Increasingly, it is very difficult for the average commoner to compete with the facilities of the modern and some intermediate homesteads, especially those in which married couples share mutual ambitions and goals. Status and prestige are coming to be based on the display of wealth, in a society where displays of material wealth have previously been frowned on.

In Chapter Four, it was suggested that the offer of cattle as collateral by some prominent Real Farmers is a subtle display of wealth. At the homestead, however, ambitious Real Farmers are able to engage legitimately in displays of wealth without offending against custom. This is seen most clearly in the example of the *lilima* (work party) which I observed at the homestead of one Real Farmer.

According to Kuper (1947), the work party was essential to the satisfaction of homestead labour needs:

> It is impossible for the homestead to satisfy its labour requirements at all times, and, apart from the assistance obtained from specialists, Swazi rely on sporadic communal work for a number of activities (1947, p.144).

Despite Kuper's claim, the work party in Ngwempisi has never been essential to the labour needs of the homestead. Community based work, either work party or *umemo* (tribute labour given to a chief or a king), is a status enhancing exercise. This becomes clear when it is considered that what are usually considered work party tasks, such as ploughing, planting, and hoeing, are carried out by all farmers at the same time in quite a competitive environment. In fact, no farmers give work parties at their busiest times, and they reacted sharply against my suggestion that they might. Work parties occur during off peak periods only. Cooperative labour does occur, particularly when women work together on tedious tasks, but this is not termed a work party, fulfilling a need for fellowship rather than labour. Most homestead requirements are met with comparative ease despite the high number of children in school, quite unlike the situation described by Kuper in 1935, when homestead labour was more plentiful.

In a work party, the Real Farmers demonstrate their enterprise through a display of wealth. In particular, they demonstrate an ability to marshal the resources necessary to host such a non-essential gathering. The work party also reinforces the public perception that the host is committed to traditional practice, complementing as it does, the practice of asking favours (*cela*). Such evidence of customary behaviour reassures kin and neighbours while allowing the host to display the resources of the homestead.

This process is well illustrated by a work party I witnessed in the off season early in September 1985 [9]. I referred to this work party in Chapter Three. The task was the shovelling of cow manure, a non-essential chore on a homestead run by a woman, who usually uses chemical fertilisers. The work party spanned a thirty-six hour period over three days, utilizing the labour of five to seven men. I made a conservative estimate of the job as needing ten man hours to complete, yet approximately ninety man hours were spent on the task. Clearly this work party was not cost effective, especially since it did not involve essential work. The guests at the work party were of two kinds. Less prominent kin supplied labour and stayed throughout the entire time. More prominent kin and neighbours visited sporadically, but did not work. The guests who passed in and out of the party numbered between fifty and sixty. All were fed and all drank large quantities of the home made *twala* (beer).

The work party was a drain on resources for no tangible return. The manure shovelled was used throughout the following season by the local children, who played on it. Eventually some was given to another Real Farmer and the rest was left largely unused.

Kuper's description of the work party implies a suppression of individuality in the interests of community, and a mechanism through which the homestead ensures it's survival. In my view, the work party is an assertion of individuality in a society which values material homogeneity. The work party represents an attempt by individuals to promote their individuality. Rather than an act of community, the work party is a manifestation of individual differentiation within a community.

The work party allows Real Farmers to act in an entrepreneurial manner, unrestrained by the control of the Chief over land and other resources. Work parties are the means by which Real Farmers are able to enhance their status without having to limit their self expression to parameters determined by the ambitions and definitions of the Chief. The work party, in its contemporary form, can be seen to be in the same category as dance or 'boasting' in prior times.[10] It is a display of resources, and is the means by which the person giving the work party perpetuates the system in which they are most accomplished.

Significantly, the work party described was given by a female manager who is also a Real Farmer. Her husband visited briefly during the work party and evidently approved. She dominated the proceedings and mobilized much support from kin and friends. I concluded that the general consensus of her guests was that she was a woman with determination, who could be relied on to fulfil her promises.

However, there are problems for the Real Farmer in displays of this kind. Any display of wealth attracts demands from kin on that wealth. This is seen when the Real Farmers are under pressure to sell vegetables more cheaply, and when they are asked to supply grain, loan money, and to provide accommodation. As I have already argued in Chapter Four, the Real Farmers are under pressure to conform to demands of this kind, though some have ways of resisting these demands.

Displays of wealth are not confined to individuals. Ntimane gave a party for some visiting maize advisers, which proved to be a conspicuous display of wealth and success. District involvement was considerable, with dignitaries, including the brother of the Chief,[11] local members of parliament, and the chairman and members of the Motane Irrigation Scheme all in attendance. As a means of confirming the status of Ntimane as the most successful scheme in that area, the party was a resounding success. As with the display at the homestead, the Ntimane display impressed visitors because the members demonstrated their comparatively considerable corporate resources, and a high degree of solidarity. These are the same factors present in displays of wealth at the homesteads, where couples exhibit their conjugal solidarity as manifested in a display of homestead resources.

Conclusion

Throughout this chapter I have argued that economic changes are stimulated by, and further encourage, significant changes in social organization. This results in a lessening of the importance of kinship in some circumstances, and in particular, changes in the nature of the family itself.

Kinship still plays an important part in the way the Real Farmers manage the affairs of Ntimane. Generally, at the homesteads of the members, changes are occurring that emphasize monogamous marriage patterns and conjugal families. Significant changes have occurred in the status of women, a situation resisted by some men, but supported by the Real Farmers who appreciate the managerial abilities of particular women. Further changes are occurring in the way young men are looking to their parents, rather than the Chief, for land and security in the future. Also, married couples are changing the way they relate to each other, favouring closer personal and work relationships. Finally, the homestead is becoming a significant arena for display, as Real Farmers demonstrate their abilities to manage and accumulate resources. In doing this, the Real Farmers are able to differentiate themselves from other people who are not Real Farmers, in a manner that makes it hard to compete for those who live in traditional homesteads, and without the resources the Real Farmers have at Ntimane.

Notes

1. In the past, selective endogamy had occurred when clans sub-divided and the new clan took the name of the head of its major lineage. This practice ceased with the consolidation of the nation, except for the

monarchs who are able to marry kinswomen otherwise excluded from them (Kuper, 1947, p.110).

2. To 'know' is to be related to them or to be intimately familiar with their lineage.
3. Kuper and Marwick, working in the 1930s, both based their homestead data on royal villages. As I have previously argued, this tends to result in over formal structures being promoted as the norm.
4. Devised to hide the grain from attackers in times of war.
5. The climax came with the death of the headman and the assumption of that status by his heir. In this circumstance the other brothers, who did not have their own homesteads, would leave the homestead taking their mother with them to form a new homestead.
6. One of the favoured stoves is a large wood stove cast in Durban (South Africa), and comprising a top, oven, and water tanks, one on each side for hot water. This stove retails at about E800 to E1200.
7. The Swaziland Proclamation No. 1 of 1916 required each polygynist to pay a tax of thirty shillings for each wife to a maximum of four pounds and ten shillings. Monogamists only paid thirty-five shillings per annum (Marwick, 1966, p.39). In 1934, in the Mbabane area 39.9 per cent of men were polygynists (Marwick, 1966, p.40), yet in 1988 less than five per cent are polygynists.
8. Lobolo is the SiZulu spelling for the SiSwati word lobola. At that time the SiZulu lexicon was used for SiSwati, despite obvious differences of pronunciation.
9. The first substantial rains began that year on the 30th October, over six weeks later.
10. Swazi have always been able to attract the attention of those higher in authority by various means, among which were dancing and 'boasting'. The former Minister of Education Dambulumjiva Hlabatsi caught the attention of King Sobhuza II with his dance after being awarded his degree from the University of Swaziland. Sobhuza was so impressed by his spirit that he sponsored his career.
11. At this time the Chief was very ill and died soon after.

6 Power and authority

Introduction

In this chapter I examine relationships between the Chief, Prince Mkhosi Dlamini, and the Real Farmers. I argue that the other bodies that exercise power in Ngwempisi do so within the constraints allowed by the Chief and in conjuction with the Real Farmers.

Traditional power and authority

Traditionally, the power and authority of a chief rested on his ability to transfer to his retainers land allocated to him by the Swazi king. In turn, the chiefs would give allegiance to the king. They provided him with political and military leadership in times of war, and exercised their judicial authority on his behalf in times of peace.

With the advent of colonialism, the power of the Swazi King to allocate land to chiefs was severely curtailed, as the colonial administration allowed expatriate settlers to take over much of the country. Judicial powers were transferred away from chiefs to colonial courts, leaving the chiefs with a Chief's Court of truncated power. However, after independence, King Sobhuza II gave his chiefs an array of powers. The King did not increase the chiefs' judicial powers, nor could he make more land available to them. However, the King did give chiefs the power to make appointments, and to approve candidates for a number of local institutions.

In Ngwempisi, power and authority are now formally exercised to varying degrees by the Chief, Prince Mkhosi Dlamini, the Rural Development Areas

Programme officials, and the Governor of the local Regional Assembly (*Indvuna ye Inkhundla*), but the power and authority of the Chief remains the most significant. The Chief exercises power in a variety of ways. He has the authority, and the symbolic power, to influence opinion directly. This power is based on commoner acceptance of the Chief's relative status, both because of his aristocratic birth, and because he is the King's representative in Ngwempisi. The Chief has direct power and authority in the spiritual affairs of his people. For example, witchcraft accusations are directed through the Chief, and he will organize a witch hunt for the benefit of his people. The Chief presides over the Chief's Court, an institution of some significance. He also controls the allocation and administration of land. Finally, he exercises direct control over the degree to which his subjects can participate in the affairs of the state.

Figure 6.1 represents the way the Chief formally exercises power. The *sikhulu* (chief) appoints a series of local officials. The *bucopho* (assistant to the Governor of the local Regional Assembly) is nominated by headmen of Ngwempisi, but approved by the Chief. He is especially important because he represents the Chief in the Regional Assembly and he also has access to local administration officials. The *umgigimi* (chief's runner) and governor are traditional positions in the chiefdom, and the functions of these offices remain essentially as they always have. The runner travels from homestead to homestead delivering messages from the Chief, and bringing people to him if he so requests. The Governor advises the Chief, and assists him in managing the affairs of the chiefdom. The *Lusendvo* (Inner Council of Kinsmen) is another traditional institution which advises the Chief more specifically on matters related to the lineage of the Chief and the Dlamini clan in Ngwempisi. The Land Management Committee is an especially important advisory body to the Chief, which deals with the allocation and use of land. The Chief nominates some of the members of this committee. The local headmen of the area elect other members subject to the approval of the Chief. The role of this committee has been discussed previously, as it is the body on which some of Ntimane's prominent Real Farmers are present.

The Chief's Court is an institution of importance, though not so much as in the past. The Chief is entitled to convene his court to deal with traditional legal matters. There are some difficulties deciding the jurisdiction of the Chief's Court (these are discussed later), but usually the court

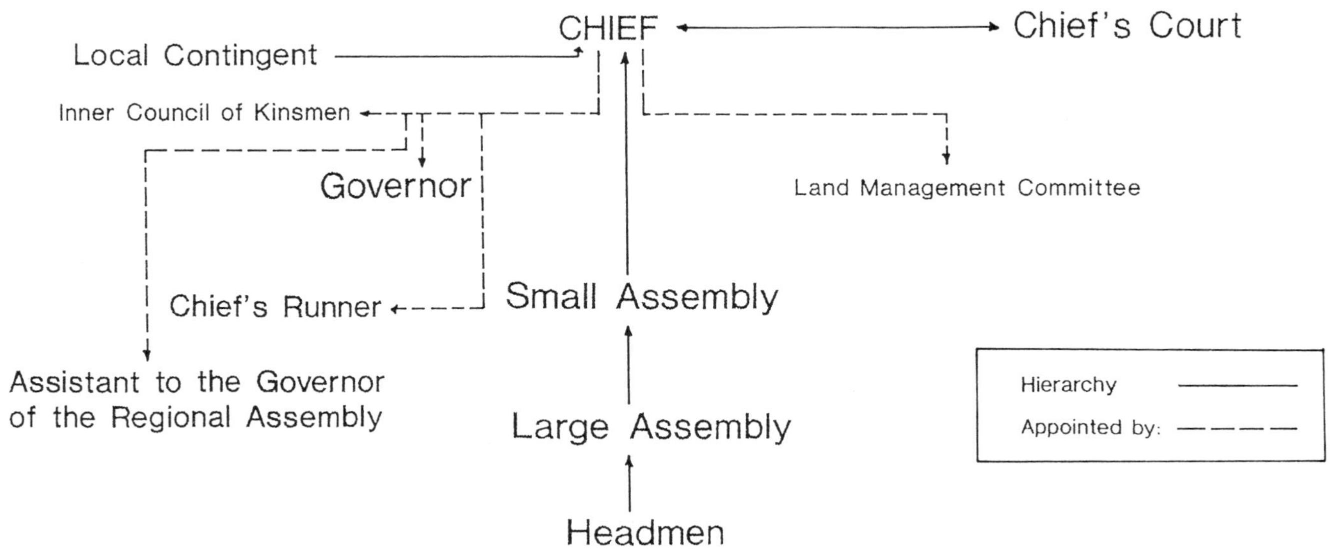

Figure 6.1 The Chief's Administration in Ngwempisi

deals with those issues which are not criminal offences, and are related to customary institutions, such as marriage cattle arrangements, land issues, and spiritual affairs.

The *Emakhanza* (Local Contingent) are elected by the people of Ngwempisi, but are candidates approved by the Chief. The Chief himself is also one of the Local Contingent. The Local Contingent, under the leadership of the Governor of the local Regional Assembly, meets in the Regional Assembly building and elects some of its members to the Electoral College.

A series of assemblies, made up of the headmen of Ngwempisi, meet to advise the Chief or to bring issues to his attention. The larger of these is the *Libandla Lelikhulu* (the Large Assembly). All headmen are entitled to attend this assembly. The Chief told me that women and young men who have not given allegiance were also allowed to attend, and to speak. However, this contradicted advice I had received from informants who said that such people had observer status only. The other assembly is the *Libandla Lencane* (Small Assembly). The Small Assembly comprises headmen elected by the Large Assembly. Once again, some uncertainty surrounds the role and composition of this body. The Chief told me that he did not influence the selection of members to this assembly. Some informants told me that he did, others were uncertain. There was consensus among informants that the Small Assembly did not meet very often and was not particularly significant in the area.

The Chief has direct access to the regional administrator based in Manzini. This right is a constituted right, and is highly significant, for the regional administrator is directly responsible to the King. This is the limit of the Chief's formal access to officials and institutions. Informally, depending on his status and closeness to the King, the Chief is able to communicate with the King or members of the royal family at will. In Prince Mkhosi's case, he is a close relative of the King and is also a prominent man in the royal Dlamini lineage. As such, he has access to the King as he needs it.

The Real Farmers in Ngwempisi are important to the Chief because they provide him with the raw material he needs to fill the appointments he makes. Because they understand the nature of rural development as it has emerged in Ngwempisi over the last twenty years, the Real Farmers provide the Chief with expert advice on matters of land use, and he is confident that they understand his desire to develop the area. The relationship between the Chief and the Real Farmers is especially strong because they support the Chief. That is, the Real Farmers provide evidence that Ngwempisi is a progressive chiefdom by running successful irrigation schemes. This evidence enhances the Chief's ability to achieve his objectives with the government officials with whom he has contact. The relationship develops because the Chief rewards his Real Farmer supporters by giving them his patronage and eventually sharing his authority with them.

The Chief's power and authority

The Chief has authority vested in him by right of birth, and through his role as representative of the King. The Chief's status is unambiguous in the minds of his retainers, and even if his real power has diminished, his authority remains strong. Such an acceptance of authority gives him considerable power to influence opinion in such a way as to achieve particular goals.

For example, although the Chief is nominally in control of Ntimane, he adopts the role of arbitrator and adjudicator, and is the ultimate authority to which the committee can turn in a crisis. Although this power is significant, the Chief exercises greater power at Ntimane because of his support for the establishment of the scheme. In return, Real Farmers at Ntimane feel a great sense of obligation to Chief Mkhosi, and they actively support him.

The Chief's influence and moral support were essential in initiating Ntimane. Since the concept was unusual for its time, the support of the Chief was essential to reassure members of the general community that the enterprise was socially acceptable.

The Chief believes implicitly in commercial development on Swazi Nation Land, and he has had substantial influence over the nature of development in his chiefdom. He told me that he believed that all people have a right to make money on their land but, that owing to land scarcity, farmers must intensify food production. He said that he believed that his support was essential for the initial success of schemes such as Ntimane, though he did feel that it is important that the committees quickly reach a semi-autonomous position of authority. In the case of Ntimane, he 'enjoys' watching his people 'progress', an attitude crucial to local acceptance of the cooperative, and one which cooperative leaders were able to propagate among both members and non-members in the area. Ultimately, this attitude also supports Real Farmer activity because all the Chief's retainers know that he favours those who show initiative as commercial farmers.

The Chief was not acting in isolation when he supported commercial development in Ngwempisi, for King Sobhuza II had attached great importance to commercial development on Swazi Nation Land, and the Chief was acting with the King's support. During King Sobhuza's reign, the Chief was close to the centre of power in Swaziland and had confidence in the strength of his position. The Chief's prominence as a member of the Swazi royal family was shown in 1973 when he acted as the 'representative' of the Swazi people in requesting King Sobhuza II to rescind the constitution and return to rule by decree.

The Chief told me that in his youth he had been contemptuous of farming as a means of making a living. However, over the years he has become certain of the need for agricultural self reliance. With the advent of the Rural Development Areas Programme, his influence was subsequently vital in questioning the argument that land and Swazi identity were inextricably linked, and instead he disseminates the view that the relationship of people to land is a pragmatic one, by telling me that:

> ... it is better to live in the countryside than in the town, for you will need food in town and you can't grow it there ... as a Swazi it is better to be in the country, for in the town you cannot even rear chickens. There is no grazing land, cattle, goats, sheep. You can't do anything ... you are just there ... in your house.

The Chief's preoccupation with agricultural self reliance is in tune with a widely held belief that relates to development in most parts of the world, and is a continuing and popular theme in Swazi politics. The crucial issue for the people of Ngwempisi is that, by propagating such an attitude to land, the Chief establishes a moral basis for differentiating access to land (see Chapter Three). Therefore, by implication, those deserving land are considered to be those who use it pragmatically.

A further aspect of the Chief's power to influence the way his people think and act is in the manner in which he expresses rural living as representing traditional cultural ideals. This notion is most pervasive, and many young people, who formerly considered city life more desirable, now refer to it. As financial circumstances change in the cities, and work is increasingly hard to obtain, so the youth respond to the suggestion that the countryside offers a superior lifestyle.

The Chief has sustained his authority by maintaining a conservative and custom oriented lifestyle. His homestead is a model of the traditional homestead, and he conducts the affairs of the chiefdom in accordance with customary etiquette. He also stresses customary behaviour, and in a major point of deviation from the Real Farmers, he argues that land scarcity is exacerbated by monogamous marriage, a situation he deplores. He suggests instead that the large homesteads associated with extended families are a more effective means of facilitating access to land than are the multitude of small nuclear families. He stated:

> Now the eldest son takes over the power of the father. So now everything will be run as if the father is alive. So the fields will be there as if the father is here; everything will be rationed. Everything will be run as if the father is here. So he [the eldest son] will be acting on behalf of his father.

By this statement the Chief meant that by supporting the eldest son as headman, he could restrict the number of new homesteads. By putting pressure on the eldest son to accommodate his brothers, as they had been accommodated while their father was alive, he could prevent large areas being taken up by the huts of new homesteads. By propagating this idea, the Chief envisages the sons and siblings of headmen sharing the homestead land, an idea not unlike the Real Farmers' intentions to allocate land to their own children. However, as illustrated by his statement to me that 'everything will be rationed', the Chief is intending to continue his personal control over the allocation of land. This is a procedure that has the advantage of retaining people within the chiefdom to prevent their need to *kuhamba bayaEuropean* (to go like a European) in the cities. The action of rationing land also has the effect of keeping the Chief central to the affairs of the chiefdom.

The interests of the Chief and the Real Farmers converge on the need for more effective land management strategies. Thus the Chief's support for the irrigation schemes is consistent with his beliefs, leading him to state: 'If they [the people] come up with schemes like Ntimane they will get enough to feed themselves' and, by implication, they would have his support. A major point of division between the Chief and the Real Farmers, however, rests on control over the allocation of land.

The Chief's use of his traditional authority to foster an appropriate moral environment for agricultural intensification is, fundamentally, a conservative action, for, in increasing the commercial viability of his retainers, he supports the conservative institutions of the chiefdom and discourages urban drift. Ultimately, he is responding to the widely held belief that a Chief is a chief because of his retainers.

The degree of authority the Chief exercises over his subjects was clearly demonstrated after 1972, when the members of Ntimane first began to deal with the personnel of the Mahlangatsha Rural Development Area. Many officials assumed that the Chief would rubber stamp their proposals for land use. His lack of enthusiasm for some proposals, specifically grazing areas involving relocation of homesteads, gave him the reputation among Rural Development Area officials of being too 'soft' on his people. The officials seemed to underestimate the degree of influence the Chief wielded, and were puzzled by the resultant public antipathy to their proposal.

I asked one government official to explain what would happen if the Chief refused to allow a land management programme to proceed. I was assured that in any contest between the ministry and the Chief, the ministry would win. The evidence does not support this contention. In fact, land management programmes, as part of the Rural Development Areas Programme, resulted in some relocations of homesteads, and some fencing, but little of substance occurred. The Chief's resistance to land management programmes did not take an active form. Instead, the Chief and his Land Management Committees moved very slowly on the issue, and stalled until such time as the officials lost interest.

The Chief is well aware of his power and authority, and stated categorically that he did not consider development a threat to his authority.

> They [the cooperative members] won't run away from my powers as they have taken the green light from me and I have the right to refuse them permission to have the scheme.

From this remark I concluded that he spoke of his power in traditional terms, suggesting that he could close down Ntimane immediately if he wanted to. In reality, this would not happen because Ntimane is so well developed, and tenure patterns so well established, that popular opposition would be too great. More importantly, such an action could not have come from Prince Mkhosi, because he recognized that the Real Farmers would never flout his authority while he exercised it appropriately, for he supports them as they support him.

By encouraging the Real Farmers the Chief creates a suitable moral climate for their operations, and so his approval continues to be a condition for all Real Farmer activity in the chiefdom. By delegating some authority to the Real Farmers, he retains the power to influence and control them. This form of patronage is accepted and expected. Kuper (1963) described this situation in the following terms:

> Rulers maintain their position by delegating authority to trusted officials, related and non-related. Nepotism, the granting of privileges to kinsmen, is an accepted principle in Swazi government (Kuper, 1963, p.31).

The relationship of the Chief with selected Real Farmers is one of trust rather than nepotism. While some Real Farmers are related to the Chief, none of the more prominent Real Farmers at Ntimane is close kin to the Chief. When the Chief delegates authority to selected Real Farmers, he does so because he knows them, and because he can trust them to support him.

To a limited degree, nepotism has played a part in influencing the membership of some people to Ntimane. In Chapter Two, I established that some of the Chief's family are members of Ntimane. Two of his wives have plots (numbers three and thirty), his son's wife is the owner of plot thirty-seven, and his brother owns plot thirty-three. The material benefit the Chief receives is limited to the produce of his wives' gardens. There is no direct material benefit for the Chief from his brother. However, the members of Ntimane told me that they were aware of distinct benefits to the scheme. They said that the connection with the Chief's family is beneficial because these members are an avenue of direct appeal to the Chief that is always open. The Chief's wives, identified to me as Real Farmers, share common goals with Ntimane's leading Real Farmers, and so have a vested interest in the Chief's continued support. The Ntimane members, therefore, are confidently aware that the Chief remains well informed of the happenings at Ntimane through informal channels, and could act to pre-empt trouble.

Another avenue of direct contact between Real Farmers and the Chief is through his daughter-in-law. Already a member of Ntimane before her marriage to the Chief's son, she is now a teacher, and because of her family connection has been allowed to use a gardener to cultivate her garden. This is a situation that is normally not tolerated at Ntimane. Because she has been allowed to operate her garden in this manner, it is generally recognized that she has incurred a considerable obligation to the members of Ntimane, which they in turn can use to appeal directly to the Chief should this be necessary.

The authority and power of the Chief are increasingly exercised in conjunction with that of the Real Farmers, which is evidenced by the degree to which prominent Real Farmers from Ntimane are involved in the higher councils of the chiefdom. The former Chairman Ndlandla, the current Chairman Ntambo, and a prominent Real Farmer Nxumalo, are all members of the Land Management Committee, which is a most important body in the local political arena. Because of the size of Ngwempisi, the Chief actually has two Land Management Committees in two parts of the chiefdom. The members are usually born in the area and are elected by the assemblies. Only 'strong' men are suitable. For example, the Chief told me:

> Ntambo is a member of the Land Management Committee, and it is seen that he is a strong man. Even if you are born in this area but are not a strong man then you will not be a member of the committee.

The men who are on these councils are able to approach the Chief relatively freely, both in the interests of the general community and on Ntimane business.

Since the members of the Land Management Committees are elected by the local assemblies, this process suggests that they are trusted by the general community. By placing such people on the committee, the general community knows that they will decide land issues in a particularly troubled atmosphere that is often emotive and potentially divisive.

The manner in which the Chief sustains some power and authority by his support of the Real Farmers is demonstrably fragile. This became evident when, soon after I left Ngwempisi in 1986, the Chief died. The Acting Chief arbitrarily ordered that no cultivation was to occur until all headmen in the chiefdom paid E100 each into a fund for a new school (Times of Swaziland, 27/10/87, 18/10/87). Since the rains had already begun, this authoritarian proclamation allowed no time for headmen to organize the payments. This action was considered to be ill advised and an abuse of power, and seriously damaged the Acting Chief's credibility. For example, prior to leaving and knowing that the Chief's health was failing, I asked prominent Real Farmers what would happen when the Chief died. They explained that nothing would change because they would advise his successor on correct procedures. The Real Farmers saw the relationship between themselves and the Chief as one of mutual trust and mutual benefit. The Acting Chief's authoritarian action is one that could alienate him from the Real Farmers, and perhaps represents an attempt to exert *kumemetela* (chiefly authority).

The exercise of spiritual power

Since so many people from the chiefdom have diversified into a variety of occupations, conflicting loyalties have emerged to undermine the centrality of traditional power. The ritual significance of the Chief remains significant, for he has the power to organize and motivate the entire community, and remains the focus of social, spiritual and psychological unity. The Chief's call on his people in special situations is almost absolute. For example, two Mliba chiefs with divergent interests met to decide responsibility for the lightning that struck a local primary school, causing death and injury. The Chiefs decided to raise E20 from each of their headmen to pay the expenses of taking their people to Lomahasha, where a witchhunt was carried out (Weekend Observer, 30/11/85). Lomahasha is close to the border of Mozambique and the *tinyanga* (traditional healers) there are renowned for their ability to smell out witches.

In this example, the Chiefs were demanding that a witchhunt be carried out in Lomahasha, which is approximately one hundred kilometres from Mliba. The Chiefs were also demanding that people leave their area during December, which is planting time and a particularly busy time of the year. When chiefs deal with spiritual matters they are particularly powerful. Another example of this power occurred in late 1986 when Prince Mkhosi gathered the people of Ngwempisi to pray for rain when the rains were late arriving. Once again all the people in the chiefdom were expected to attend, regardless of their prior commitments, and in many cases, rural wage and salary earners returned to the chiefdom for the event.

The single most important issue the Chief handles is that of witchcraft. All accusations of witchcraft are directed through the Chief, and theoretically, because of his intervention, the spiritual well being of the people is protected. The sense of spiritual wellbeing rests on a general acceptance among people that the Chief himself has not been corrupted by sorcery. For example, Kuper (1963) stated:

> Important men do not need to use sorcery against insignificant inferiors, nor are they suspected of doing so. Sorcery is an indication of status and of the ambitions for improvement of status that operate within the limits of the stratified traditional society (Kuper, 1963, p.66).

Commoners accept that the chief's status is far above their own, and so do not contemplate using witchcraft against him. Because they are so aware of his status superiority, they also do not conceive that he might be jealous of them and so do not imagine that he would use witchcraft against them. Fear of jealousy, leading to witchcraft, is intense among commoners, and as Kuper suggests, it is fear of witchcraft that forces individuals to act carefully so as to avoid the jealousy of others. Therefore theoretically, through the impartial adjudication of the chief, fundamental social stability is ensured.

Clearly, the Real Farmers are highly dependent on the support of the Chief in this area for, as Kuper states, status and ambition are a focus for jealousy in this stratified society. Their success rests on their ability to bring about change without incurring the enmity of their neighbours and kin. By cementing a firm and mutually rewarding relationship with the Chief, the Real Farmers are able to ensure that they have the protection they need from witchcraft, in an environment in which suspicion of their actions is not only unrewarded, but can also evoke sanctions.

The traditional manner of dealing with witchcraft was to hold a witch hunt, which has happened in Ngwempisi in the past. In the early 1970s, after the unexpected death of a prominent headman, Prince Mkhosi called a witch hunt in Ngwempisi. Unlike the Mliba chiefs, he brought the witch hunters to Ngwempisi for a ceremony which involved the entire community. In this ceremony seven wizards were smelt out.

People stated that the witch hunt was necessary as jealousy had been getting out of hand, and people had been suffering. It is evident that the emotional and psychological release after a witch hunt is profound. For those accused, the stigma of being a recognized wizard is with them until they die, although in contemporary Swaziland they survive their unmasking. In the past, immediate execution had always followed discovery. The means by which the general population copes with the wizards in its midst is to: '... watch them and never look them in the eyes'.

The successful witch hunt enhanced the Chief's ritual power. It is also significant that, of those smelt out, none was an important Real Farmer, though accusations were made. The witch hunt not only bolstered the Chief's role, it reinforced the activities of the Real Farmers at a particularly stressful time, when the resources of the Rural Development Areas Programme were flowing into the chiefdom for the first time, and people became jealous of the comparative advantages of those established schemes in gaining resources.

Despite the gains the Chief made from the witch hunt, public ceremonies of this kind do endanger the Chief's control over the situation. Most witchcraft accusations never reach the stage of a formal ceremony, for once that occurs, control over the situation moves into the hands of the witch hunters and away from the Chief. It appears that the Chief preferred to adjudicate such matters himself if possible, to achieve his own ends which, in the examples I heard of, were to support Real Farmer activity. In these examples the Chief intervened to diffuse any potential divisiveness, and maintained his personal control over the situation.

In the first example, a young woman, now a prominent Real Farmer (see Chapter Three, Case Study Two 'Intermediate Homestead'), married into an important commoner household. Because of her commercial initiative, she incurred the jealousy of her husband's sisters. Subsequently, when her father-in-law died, she was accused of witchcraft by her sisters-in-law, an accusation she publicly challenged. The accusation against the woman was an attempt at character assassination, which received no support from the Chief. The young woman, in challenging publicly and boldly, forced a retraction of the accusation by the other women. The Chief played a significant part in the affair by not allowing the matter to reach a formal hearing.

The second example involved witchcraft directed against a prominent Real Farmer couple, in which the wife almost died. A traditional healer was able to cure her and accused a male relative (and fellow member of Ntimane) of witchcraft. The husband took the accusation to the Chief, who confronted the male relative. This man neither admitted nor denied the charge, but the exposure before the Chief had a powerful sanctioning effect. By chastizing the man, the Chief diffused the situation without bringing it into the public arena through the use of a witch hunt.

The Chief is not alone in using witchcraft allegations to achieve particular goals. The Real Farmers, too, have used witchcraft allegations to maintain their control in Ntimane. I have previously referred to one such incident, when the Real Farmers used an accusation to justify evicting a member from his garden (see Chapter Two 'The Real Farmers at Ntimane'), having already seriously considered ejecting him for other infringements.

In another example, an early member of the scheme, Msibi, constantly fails to enter into the dynamics of farming at Ntimane, being an unwilling member and a poor farmer. Over the years his lack of cooperation has rankled the leading farmers and he is often spoken about in a derogatory manner. In particular, he was slandered by reference to him as a 'wizard', an accusation successfully levelled against his brother during an earlier witch hunt. A clear trend was emerging, whereby the Real Farmers were building up general resentment against this man to justify taking his garden away from him.

Fear of witchcraft remains a potent force for social control in Swaziland and, since the zionist christian church is a syncretic church, christianity has reduced neither the potency nor the incidence of witchcraft. The stabilizing role of the Chief is, in many ways, more necessary now than ever. But even chiefs have been caught up in the competitive environment of change, and some have allegedly also been involved in witchcraft. This is not a new phenomenon and seems to be related to a form of social stress.[1]

Accusations of witchcraft against a chief are a terrible blow to public security. These accusations seem to occur in situations where chiefs have less power, which they then seek to enhance by resorting to witchcraft. For example, in Northern Hhohho, a heavily populated area near the Komati river, ritual murder is allegedly very common. A recent rash of killings has resulted in allegations that the Chief of that area, or people close to him, are involved (Times of Swaziland, 20/11/85, 26/11/85, 27/11/85). The reports suggest that the murder and mutilation of one seven year old boy for ritual medicine was intended '... to strengthen a certain man in his position.' (Times of Swaziland 26/11/85). The delicacy of the insinuation, and subsequent unsubstantiated reports, suggested that Chief was closely associated with the incident.

Ritual killings are said to occur constantly in over populated areas such as Hhohho, or where populations are under stress (Booth, 1983, p.50). The response from the highest levels of government is one of abhorrence, for such activity by a chief diminishes the spiritual integrity of the royal clan. Prince Mancibane, a member of the *Liqoqo* (Inner Council) and a prominent royal family member, responded on behalf of the government, by calling for the public burning of ritual murderers (Times of Swaziland, 9/12/85).

Social stress is a well recognized factor in the incidence of ritual murder in Southern Africa.

> In Basotholand, [Lesotho] since the death of Chief Griffith in 1939 and then of his recognized successor Seeiso in 1941, there had been a pathological eruption of murders of innocent people whose bodies were mutilated for medicines. The guilty included leading men, and the situation could be linked with the legislation that had limited the number of chiefs recognized by government and given effective power. The case of Chief Bereng, a son of Griffith, was particularly disturbing. Bereng considered himself the rightful heir and, after the death of his brother Seeiso, challenged in the courts the appointment of Seeiso's widow Mantsebo as Regent. He lost the case in 1949, was found guilty of complicity in a medicine murder case and executed (Kuper, 1978, p.176).

During the interregnum of Prince Mfanasibili (1982-1986), as individuals jostled for power, ritual murder seemed to be more common. Whilst this is unsubstantiated, public interest in ritual murder certainly increased remarkably, with constant newspaper reports of cases, and accusation and insinuations being common. Even Prince Mfanasibili himself was popularly alleged to be a wizard and, it is rumoured, had tried to steal items from the remains of King Sobhuza II to enhance his own powers.

In Ngwempisi, as population levels have not reached the critical levels they have in other parts of Swaziland, ritual murder has not reached significant proportions. The people of Ngwempisi are very aware of the occurrence of such events in other places, and are also aware of the comparative peace of their own chiefdom. Nevertheless, witchcraft is significant in Ngwempisi, and is still an arena in which accusations and counter accusations play an important part in power relations.

The Chief's Court

The Chief's Court is another avenue through which the chief exercises significant power. Swaziland retains the old colonial dual legal system of civil and traditional courts, through which, in some cases, litigants may exercise a choice of court (Kuper, 1963, p.37). Armstrong and Nhlapo (1985) describe this in the following terms:

> Swaziland thus has a classical dual system, a state of affairs that is typical of most African countries with a colonial past. The 'received' law is administered by the courts of general jurisdiction (the subordinate Courts and the High Court), while the Swazi Courts are restricted only to the application of customary law (1985, p.3).

The chief arbitrates in many other areas of traditional law, beyond that of witchcraft. Except for criminal cases, the litigants can choose their court, depending on how traditional the issue is considered to be. The choice of the Chief's Court is considered popular for solving a variety of issues. Significantly, civil courts are preferred for resolving financial matters. In these latter situations commoners are sufficiently astute to prefer personal distance and legal formality, as opposed to the intimacy of the Chief's Court, where every nuance of litigation becomes general knowledge within the chiefdom. Since litigation over money usually involves potentially embarrassing disputes among kin, many prefer to take the issue out of the chiefdom.

The advantage of the Chief's Court over civil courts is considered to rest in the chief's intimate knowledge of local affairs, ensuring an appropriate judgment. The Chief's Courts are not dehumanizing, and are conveniently located. People also know the chief and the venue (the chief's homestead), they are familiar with those who are attending, and they and are conversant with the procedures involved. Civil courts have long waiting lists, are far away, and are impersonal. Commoners, however, seem to enjoy the paternalism of their contact with the legal process through the Chief's Court.

In seeking justice they may often feel that the chief has the requisite local knowledge and spiritual and moral authority to deal properly with a problem. For example, I earlier referred to the situation where a man was 'given' land by his maternal grandmother, a gift subsequently disputed by his mother's male kinsmen (see Chapter Three, Case Study Three). In this case, the man was accommodated on land provided by his Real Farmer father-in-law. The matter went before the Chief's Court, which judged in favour of the maternal relatives, but approved the moving of the homestead only some 100 metres. The maternal relatives failed to use the land, and the man has since recommenced using it with the Chief's approval. So insubstantial was the move suggested by the Chief that the farmer retained easy access to the fields. I concluded that the Chief recognized the legitimacy of the relatives' claim, but knew the land would not be used. By recommending a minor move of homestead, he satisfied the relatives without actually preventing convenient use of the land. When I asked the headman about the affair he was non-committal, telling me only that he was pleased with the Chief's judgment because the Chief '... knows what is required'.

If litigants are sure of their case and have strong local support or enjoy the patronage of the chief, they are reluctant go to a civil court for fear that the impartiality of justice may negate the advantage they already have. Furthermore, such an action might offend the chief and prominent patrons by questioning their ability to resolve the relevant issue.

The fundamental issue involved with the Chief's Court is the basis on which people choose to use this court or another. Clearly, choosing to use the Chief's Court will either be the action of an individual sure of his standing with the chief, or of the chief's judicial integrity. The Real Farmers enthusiastically support the right of the Chief to adjudicate through the Chief's Court. Since the Chief is progressive and supports the Real Farmers, they can expect a sympathetic hearing. Surprisingly, the incidence of litigation in the Chief's Court in Ngwempisi is low, as the Chief prefers to settle all such affairs in a more informal manner. This strategy rests more on public acceptance of his authority than on the application of his power. So complete is public acceptance of the Chief's right to adjudicate that, like matters of witchcraft, judgements made outside the formality of the Chief's Court are not disputed by commoners. At no point are people prepared directly to confront the judicial authority of the Chief, despite the fact that, in real terms, his ability to use severe sanction is limited.

Central government power in Ngwempisi

The chief interacts with the various arms of government at three levels, and within the traditional and modern spheres of formal government. At the first level, he controls the ability of his people to participate in the mechanism of government through the Regional Assembly system, and through the Swazi National Assembly. At the second level, he has direct access to government officials at a high level. The final level is that of his personal contacts. As a member of the Swazi royal clan, and close kin of the King, he is in a position to approach significant figures informally to achieve a particular result.

The Regional Assembly system and the Swazi National Assembly The chief influences the selection of candidates to the local Regional Assembly, though the approved candidates are said to be democratically elected.[2]

> Voting at Tinkhundla [Regional Assembly] level is not by secret ballot but are *[sic]* conducted in accordance with traditional customs. For instance, each inkhundla's executive [*bucopho*] suggests the names of four persons and out of the four, the two with the highest votes become the delegate[s] of that particular inkhundla in parliamentary elections. The candidates are not consulted nor is the Chief told of the candidates until they are to be elected (Shongwe, 1983, p.20)

Shongwe's claim that neither the Chief nor the candidates is consulted is dubious and unsubstantiated. Indeed, he later contradicts himself by claiming that the Regional Assembly system concentrates power with the '... princes, chiefs and some relatives of the *Ngwenyama*' (1983, p.36). Certainly, in Ngwempisi, candidates are approved by the Chief through the assistant to the Governor of the Regional Assembly. It is from the Regional Assembly that the Electoral College and, subsequently, the *Indlu Ya Bayasincane* (House of Representatives) are formed. The Chief also exercises direct influence over the Governor of the Regional Assembly through the assistant to the Governor of the local Regional Assembly, whom he appoints. Problems have emerged, however, for the relationship between the Regional Assemblies and the chiefs, with some chiefs complaining that poorly sited Regional Assemblies have undermined collective decision making processes in the chiefdoms (Shongwe, 1983, p.34).

Chiefs have also reacted against the entitlement of the local governor of the Regional Assembly to make calls on the chief for assistance and for labour. Chiefs always had the right to call together labour, and such tribute labour confirmed the status of the chief concerned. The granting of rights to the governor of the local Regional Assembly to also call together labour was seen by many chiefs as a threat to their traditional rights. The potential for conflict was further compounded by governors also having the power to summon the chiefs themselves (Shongwe, 1983, p.7).

From the outset of the Regional Assemblies system, even in the pre-independence years, poor relations with the chiefs were a problem:

> ... in most cases tindvuna summon meetings and the chiefs are expected to attend without any advice on the agenda. Hence chiefs tend to regard Regional Assemblies with suspicion (Shongwe, 1983, pp.31-34).

This antipathy remains a cause for concern at a national level. In the media, or in parliament, chiefs are either blamed for being 'non-cooperative' (Times of Swaziland, 19/2/86), or they are ostentatiously praised for their cooperation (Times of Swaziland, 12/5/86). In these two cases, inferences were made that the main failing in the Regional Assemblies system was because chiefs resented being 'summoned' by the governor of their local Regional Assembly. The conflict between the chiefs and the governors of the Regional Assemblies demonstrates how the extent of the chiefs' cooperation can disrupt the workings of the Regional Assemblies. This power held by chiefs is cause for resentment and uncertainty among the governors.

In Ngwempisi, the general perception that the Regional Assemblies are in some way inadequate is compounded by confusion over their role in the political process. This lack of a clear role delineation was one of the early faults discovered in the system, and could be the result of King Sobhuza II's failure to define the function of the Regional Assembly system when he launched it in 1953 (Shongwe, 1983, pp.31-32). Confusion over this issue became evident most recently when, on 5 November 1987, the nation went to the polls to elect the members of the Electoral College, who would in turn elect the Houses of Parliament. The election proved most controversial, was poorly attended (Swazi News, 7/11/87), and ended in the alleged rigging of the election of senators to the Upper House (Times of Swaziland, 24/11/87, 25/11/87). Dissatisfaction with the electoral process was evident before the election, when it became clear that all major decisions had already been made, and that few people understood the electoral process.

A public holiday for election day, to facilitate the election, was announced only one day prior to the event. Such was the confusion on election day that even the organizers of the election were not sure of procedures (Times of Swaziland 4/11/87). Letters of complaint were published (Times of Swaziland 12/11/87), and even the Governor of the Regional Assemblies, the Great Councillor, Mr Mndeni Shabalala, stated that the final rigged election of senators '... was irregular and had been a major fiasco' (Times of Swaziland 24/11/87).

The Swazi National Assembly is also cause for some confusion, and is an area of significant control by the chiefs over their retainers, as the chiefs approve those wishing to participate. The confusion occurs because of ambiguous understandings among people as to who exactly is allowed to attend the Swazi National Assembly. The Chief told me that all men and women have the right, not only to attend, but also to speak at, the Assembly. I was told by other informants that, while all could attend, only men could speak. Other informants argued that only men approved by the Chief could attend, or that only headmen could attend. Most informants agreed that they would not speak at the Assembly unless they had the Chief's prior approval. This uncertainty over eligibility to attend the Swazi National Assembly is at variance with Kuper's (1947) description:

> Its [the Swazi National Assembly] members include, firstly, all the chiefs and their leading counsellors; secondly prominent headmen; and thirdly, any adult male who wishes to attend (Kuper, 1947, p.62)

The King calls together the Swazi National Assembly on an irregular basis, but it is not an expedient way of resolving issues for a variety of reasons. First, uncertainty over the rights of commoners at the National Assembly effectively ensures that leading chiefs and governors will dominate as advocates for the people. Second, the King usually calls the Swazi National Assembly when there is a specific issue of significance to be addressed. In these instances he has an agenda, while the public has diverse views on what is topical. Third, commoners are most unwilling to stand and address the King, and would be most reluctant to do so in such a forum. Informants told me that to address the King in such a direct manner would be a frightening experience because it contradicts traditional notions of how to show respect for someone as exalted as the King.

This latter point is well illustrated by a story I heard from one man concerning the period of unrest just after Independence. On one occasion two men were accused of an offence, and decided to summon the King to court. They persisted, despite great public dismay, and the King agreed to come. The first man, when he rose to question the King in the court room, lowered his eyes and sank to his knees in the traditional manner, and begged the King to forgive him for his audacity and lack of respect. The second man boldly questioned the King, and looked directly at the King's face, to the dismay and the disgust of the viewing public. The King answered his questions and then left the court. The first man was cleared of the charges, while the second was duly charged and punished. My informant told me that punishment was necessary because the man had committed a grievous offence by challenging the King. When I suggested that the punishment was for the original offence, the man assured me that the court would have been more concerned to punish the man for his disrespect.

In this example, whether the man was punished for the offence or for disrespect, my informant clearly believed that the legal process was subordinate to traditional obedience and respect for the King. This was a common view and demonstrates exactly how difficult it would be for a commoner to address the King at the Swazi National Assembly on any contentious issue.

The Swazi National Assembly is nevertheless significant for commoners. The general consensus of opinion that I encountered indicated that the Chief has much to gain by having a significant presence at the capital whenever there is an assembly. To achieve this, he requires a substantial group of people from his chiefdom to accompany him. Additionally, he must have retainers with him who support him and show enthusiasm for the opinions he makes public. I concluded that, in such circumstances, the Real People would constitute an extremely important collection of well disposed retainers, whose support the Chief needs constantly to cultivate. Among the Real People, the Real Farmers are especially important to the Chief in these circumstances, because they are demonstrably commercially successful, thus enhancing the Chief, but still reliant on him for their success.

The government bureaucracy Despite the limitations evident in commoner access to various levels of political representation through the role of the chief, he continues to be highly regarded, and commoners accept his mediation. The status gained by the chief in his 'traditional' authority is considered reasonable remuneration for his vitally important ability to gain access to bureaucrats at many levels within the administrative structure of the Swazi state. This is especially recognized by the Real Farmers, one of whom told me that the Chief is '... most important to me personally, and as a member, for everything goes through the Chief.' The current chairman of Ntimane told me forcefully that the Chief is most important to Ntimane '...because the ministers [of government] report to the Chief, and then the Chief takes it to the people'.

Though it is anticipated that the farmers would deal with the bureaucracy through the appropriate channels, they soon found this to be limiting. Prior to the advent of the Rural Development Areas Programme, the Chief had ' ... sole charge of his area and his people ... [and was] ... directly responsible to the King-in-Council for all administrative matters affecting his area' (Government of Swaziland, 1978a, pp.19-21). As previously discussed, the introduction of the Regional Assemblies has had limited impact on the way the Chief exercises power. However, the introduction of the Rural Development Areas Programme did threaten his authority, by giving real power to those administering the projects.

Relations between the people of Ngwempisi and the bureaucracy operate on two levels. The farmers have personal contact with the rural based bureaucrats, while the Chief is able to deal with the urban based administrators. After the Rural Development Areas Programme was initiated, problems arose over the self perception of bureaucrats, some of whom assumed many of the social characteristics appropriate to those with authority. These characteristics the rural commoners interpreted as arrogance, which, although regarded as appropriate among aristocrats, is offensive among commoners. The predominantly young men of the Rural Development Areas Programme failed to understand that they have authority only while they have the support of the Chief. In alienating the Chief, they lost this support. This loss was further accentuated by their relative inability to deliver on the programme promises they made.

The rural based bureaucrats were soon seen to be unreliable, and increasingly the farmers fell back on the Chief's right to approach directly the Regional Commissioner, who is head of the Regional Council. This council is the point through which the entire Manzini District is administered. As a result, direct access to the Regional Commissioner is potentially very rewarding for the Chief, but serious weaknesses lie in the potential for poor relationships to develop between the administrator and the chiefs. Some chiefs resent the intrusion of the Rural Development Areas Programme, and some administrators are arrogant and unapproachable. An example of this latter point was an episode at Ntimane when a district official arranged to meet the members of Ntimane at a particular time in the morning on a certain day. The members gathered at that time, and refreshments were provided in anticipation of the official's arrival. The farmers waited for about four hours until mid afternoon, and then decided that the official was not coming. Having made this decision they recommenced work in their gardens. Late in the afternoon the official arrived at Ntimane. He was obviously very drunk and became annoyed when he found that the farmers were not still waiting for him. He started abusing the farmers, telling them that they had no respect for him and that the business which had delayed him in Mliba was more important than his business with the members of Ntimane. The farmers provided refreshment, which he demanded, listened to his tirade and waited for him to leave.

This episode was significant in that when the official demanded that the farmers meet with him, the most prominent Real Farmers refused. They stayed in their gardens and sent other members to attend him. When the official left, all the farmers gathered. They were extremely angry and embarrassed, and discussed the episode well into the evening. I was told by several Real Farmers that the official was very rude and was behaving badly. One man said to me ' ... does this man think he is a chief?'

I concluded that, by staying away, the prominent Real Farmers avoided having to suffer the indignity of dealing with the drunk, arrogant official. At the same time, by sending other members to serve him, they were being careful to placate him, relying on his drunkenness to blur his memory of the events of the day, for, in the final analysis, officials are powerful, a fact the Real Farmers recognize.

There is a point at which chiefs and the senior bureaucrats have much in common. In seeking to enhance their status in the eyes of the King, bureaucrats require evidence of successful administration, and the chiefs require some control over funding and manpower. On some occasions administrators have acted without the concurrence of the chiefs, and some chiefs have then undermined development work in their area. The conflict between chiefs and administrators proves unequal, since the chief has the added advantage of direct contact with the King through traditional channels of communication, while regional administrators can only communicate with the King through the Prime Minister.

The interests of the producers and the government workers differ since the latter are not dependent on the local economy. The officials, unlike the Chief, do not depend on the success of commercial farming in Ngwempisi for their living. They are usually comparatively young, urban orientated, and politically naive. Their youth is often seen by older people to militate against them, because they are seen as relatively unsophisticated and socially unlearned.

Farmers are particularly concerned by the relative lack of familiarity the government officials have with the real issues of the chiefdom, such as the complex social issues constraining individual rights over land. The Real Farmers see this as a lack of empathy with the chiefdom, and several prominent Real Farmers made explicit references to the need to get local youths into the government system, and then back into the area as technical experts. They told me that this would offset what they see as a lack of accountability on the part of government officials, because they would 'know' the official and would be able to influence him.

The Real Farmers have also set out to cultivate some government officials. For example, the field officer for the Malamba Rural Development Area is respected and liked by the members of Ntimane. Despite the fact that he is constrained by a ineffectual organization in the service he can offer, he is interested in the Ntimane farmers and maintains contact with them. The farmers see this contact as valuable in that he can provide them with advice and insights into the politics of the Rural Development Areas Programme. In return, the young official has made good contacts in the area, he has come to know the Chief well, and he shares some of the kudos for the success of Ntimane. This latter point is interesting because this official is not in any way responsible for the success of Ntimane. Nevertheless, he has used his contacts there to obtain permission to bring guests to Ntimane. In this way he has used the scheme as advertizing for the Malamba Rural Development Area, and for his own purposes.

By helping the Chief maintain his authority despite the Rural Development Areas Programme, the Real Farmers have enhanced the opportunities presented by the programme. That is, they have prevented the officials from gaining power by circumventing them, yet they have continued to gain access to the resources offered by the programme. This relationship with the Real Farmers has diffused the Chief's authority and power. At Ntimane, prominent Real Farmers have assumed areas of authority that were formerly the prerogative of the Chief. For example, recently the Chief decided to allow the Ntimane committee to make all final decisions over the running of the scheme. Until that time all disputes have been adjudicated by the Chief. In this instance, the Chief has placed his judicial authority and his absolute and recognized right to exert chiefly authority in the hands of commoner individuals. This transference of his authority is limited to particular Real Farmers whom he considers sufficiently competent to cope with the responsibility.

By delegating power and authority in this manner, the Chief reveals new avenues by which he gains power and authority. He moves away from his traditional control over land, as the basis of power, and establishes himself as the appropriate source of legitimacy and the final arbiter of tradition in the chiefdom. Pragmatic managers recognize this. For example, the manager of part of the huge wood pulp forestry operation, which adjoins Ngwempisi, told me that he maintained a good relationship with the Chief, and always sought his approval prior to instituting any new operation in the adjoining area, since the Chief's goodwill is essential for a harmonious workforce.

The Real Farmers recognize the continuing importance of the Chief's role in maintaining stability in the chiefdom. They reinforce the Chief's position by stressing the importance of tribute labour, and by involving themselves in community work. One example of this was when the Ntimane farmers joined in the giving of tribute labour in 1985, and brought their tractor with them. This allowed for the early and deep tilling of the Chief's fields with a resultant good yield. Another example was that of the school building project which the Chief had endorsed, where Real Farmers, and Real People generally, not only donated the requisite E100 per headman, but also carried out much of the manual work in well organized work parties. By actions of this nature, the Real Farmers confirm the Chief's importance to themselves. Similarly, the Chief receives the credit for a well run, well motivated chiefdom.

However, it is significant that some individuals in the chiefdom challenge the Chief's right to call for tribute labour. One instance occurred during my time in Ngwempisi when a man refused to report to the Chief's homestead as demanded. This case was not resolved because the Chief died soon after. Young men who have not given allegiance are also blase about the Chief's role in the area. Several young men told me that they felt that the Chief was superfluous to their future, and that he exercised too much power in the area. However, I was interested to note that on the occasion of the Chief's visit to Ntimane, these same young men acted in a highly deferential and respectful manner towards the Chief.

The Rural Development Areas Programme was a catalyst for great changes, but these changes were largely unanticipated. The inevitable consequence of the programme is that it fed the opportunism of the Real Farmers, and reinforced its own inadequacy by providing the Real Farmers with enough broken promises to ensure constant criticism. This allowed the Real Farmers, and certainly the Chief, to benefit from the disillusionment that subsequently occurred as incipient commercial farmers impatiently awaited promised services.

Even the most astute Real Farmers found the problems of dealing with the bureaucracy taxing. In the words of one of them:

> I am not happy [with the government] because when we report our problems to the government agencies they are not fast in getting action. Meanwhile we are losing because we are sitting here waiting for something to happen. When the government takes a long time processing the thing [the complaint] it doesn't impress me ... it is the same problem [with the Rural Development Areas Programme and irrigation department] because it takes a long time to do a small thing.

Major misunderstandings often occur over the manner in which government personnel present themselves to farmers. The fieldworkers for the Ministry of Agriculture and Cooperatives routinely refer to their activities as those of the 'government'. This has developed into the perception among farmers that the officials represent a 'government', in some way distinct from the 'government' they know, which is represented by the chiefs and the King.

I concluded, from the manner in which the farmers talked of government officials, that they did not see a strong connection between the King and the chiefs, and the Prime Minister and ministers. Rather, they saw a second government; one made up of bureaucrats, with money and resources, but without much political credibility. Therefore, while they reacted to initiatives from 'government', they did not trust the officials.

Ordinary farmers have such difficulty delineating between the authority of the government, as represented by the ministries, and that of the chiefs and the King that they have utilized the Chief as arbiter, supporting the bureaucratic decisions that he supports and rejecting those about which he expresses reservations. Prominent Real Farmers actually seem to have a good understanding of the role of the government officials, but general public confusion over the issue is to their advantage, representing a further aspect of knowledge as power.

The farmers have become wary of the government, measuring the worth of the central government by the promises of its employees. Bureaucratic credibility has become so strained by incessant ineptness and corruption that open antagonism towards government employees has been shown. For example, at Ntimane a badly needed pipeline, with a planned construction period of three months, commenced in 1983 and finished in 1988. On this project the farmers have organized their own finance (AD 20,000) and labour, but were obliged to turn to the Irrigation Department for expert advice and assistance. This assistance was marked by the most appalling incompetence and duplicity, at extreme inconvenience to the farmers. Their relationship with the Irrigation Department is a litany of countermeasures. They were advised to buy large quantities of cement (worth E10,000), which they then had to sell to avoid spoiling after a year of waiting. The official supervizing the project was incapable of fulfilling his duties and attended only sporadically. When the farmers complained, they were accused of laziness, and told that they, rather than the official, had failed to appear, an accusation that was transparently false. In many small ways the Irrigation Department neglected the farmers. For example, one small component needed to be cut and welded; it was a simple chore and although the job time was estimated at one week, it took three months to complete. The farmers were so disillusioned by their experiences with the Irrigation Department that they have since deliberately pursued contacts in private enterprise as a way of getting the services they need.

The experiences the farmers have had with government officials has reinforced their relationships with the Chief, and within their community. The rapid growth of the Real Farmers, as people wielding legitimate power and authority, can be credited as much to disillusionment with government as to their own opportunism.

The world view of the farmers is expanding rapidly, incorporating foreign trade (since they sell to South African middlemen) and a growing awareness of their position in the Swazi state. This is a situation illustrated by the degree of unease they felt over the interregnum of the corrupt Prince Mfanasibili. Their unease went beyond concern over the succession rights of Prince Makhositive (now King Mswati III). It was also based on their experience during that period with corruption, which threatened their supplies of seed and fertiliser from South Africa, owing to corrupt licence deals.

Ironically, the farmers of Ntimane and more generally, Ngwempisi, increasingly operate in circumstances decided on, and controlled by, the same central government they abjure. The mundane realities of their economic existence are forcing a greater dependence on the infrastructure provided by a central state. They rely heavily on the movement of middlemen, who transport their goods to markets in the major centres of Swaziland, and into South Africa. Similarly, they rely to an increased degree on petro-chemical inputs and hybrid grains which come from South Africa. Furthermore, many farmers are capitalizing rural operations with wages remitted from the towns or from the large estates. Increasingly, selective market forces require farmers to be more consistent in the quality and quantity of their produce, with the result that the use of cold storage, all weather roads, and milling facilities are inevitable. This will necessitate linkages to major utilities, and with such linkages come taxation and rates.

While accepting thc inevitable, the farmers exercise extreme caution and some misgivings, for they are aware of their growing dependence on central government. The advantages of diminishing bureaucratic influence at the local level lie in the farmers' relative economic freedom.

However, the farmers, under the guidance of the Real People and the Real Farmers, are cautious and increasingly astute operators, with a mistrust born of experience. In the long term, this generation of mistrust may prove to be the Rural Development Areas Programme's greatest contribution to the development of commercial agriculture on Swazi Nation Land.

Conclusion

To sum up, the confusion in the 1970s and 1980s concerning overlapping areas of authority and power within the chiefdom has largely disappeared. Local power now rests with the Chief and with the Real Farmers, all incorporated within the general category of Real People. This relationship may eventually prove unequal as the Real Farmers continue to acquire more power.

There are great opportunities available for chiefs who support the King through exercise of their authority in such a way that their chiefdoms support the political authority of the King. Competition for the King's favour is increasing, and huge resources are administered through the Tibiyo Taka Ngwane Corporation, which is owned by the royal family. Successful chiefs are those who deliver a unified chiefdom to the King. In return for this, benefits and rewards from the central government are directed through them back to their chiefdoms.

The relationship between the Real Farmers and the Chief in Ngwempisi will remain important while Swaziland remains a monarchical state. The years of rapid growth and change may well be over for Ntimane, but I believe that the emergence of the Real Farmers has only just begun. By gaining, and then sustaining, the support of the Chief, the Real People achieve respectability, a moral basis for their actions, and some degree of protection from witchcraft and the excesses of bureaucracy. Legitimacy for their actions will continue to be important, but then they have always known how to reinforce tradition to facilitate change over the long term. For example, in 1986 Real Farmer influence in Ngwempisi gained new significance. The Chief died, leaving a child as his heir. Usually an interregnum is a time of uncertainty, but the Real Farmers at Ntimane are certain of the future, for the child's mother is a member of Ntimane. She is one of the women who has gained proprietorial rights over land on Swazi Nation Land for the first time in Swazi history. The interests of the Ntimane Irrigation Scheme, and her personal interests through her 'house', will inevitably match those of the next hereditary chief of Ngwempisi.

Notes

1. An account of 'Mamisa the Warrior', by Alister Miller (1933), recounted tales of ritual murder in Hhohho at the turn of the century (1933, pp.9-10). The story of Mamisa is set in the time when the pressures of the concessions period and the instability of political leadership were causing a great deal of social stress within Swaziland.
2. The chosen candidates are each seated in a gate through which those voting for them can pass. Although choice exists, it is also possible for all to see who is voting for whom.

7 Social differentiation

Introduction

In Chapter One I suggested that contemporary Swazi society is more differentiated than traditional Swazi society. In this chapter I will describe the form social differentiation has taken in Swaziland as processes of production have changed, and how this affects differentiation in Ngwempisi. The emergence of new levels of differentiation involves a dynamic process in which some individuals perceive opportunities and gain the means to acquire a more significant position within the hierarchy, while others experience and perceive such developments as creating dangers of declining position.

Social differentiation

Since the accession of King Mswati III in 1986, royal power has continued to be sustained and enhanced by the royal family's control over state companies and the political process. At the local level the power of the chiefs has diminished proportionate to their ability to adapt, but by gaining Real People support, considerable influence and power remain open to them. So, whilst royal power has increased, a diffusion of power has occurred at the local level, bringing with it social differentiation which is expressed increasingly in material terms and is based on individual initiative, without reliance on clan and lineage status. As a result, differentiation can be discerned via an examination of five categories of people; the ruling aristocracy, the urban elite, wage and salary earners, rural commoners, and the Real People. The

Real People, and their sub-category the Real Farmers, have been discussed in Chapter One.

The ruling aristocracy and the rural commoners are categories that equate with pre-existing social categories described by Kuper (1947, pp.54-71) as 'aristocrats' and 'commoners'. The aristocrats were members of the Dlamini clan from lineages close to the king, while the commoners were those who were not royal Dlamini, even if of important lineages in significant clans. The urban elite and the sub-category of local government personnel are identifiable because of their status as public servants, but the designations are mine. The categories of the wage and salary earners and the sub-category of the Chief and headmen are categories I have identified and named. The Real People and the sub-category Real Farmers, are categories identified to me by informants.

The ruling aristocracy

The ruling aristocracy predominantly comprises members of the senior lineages within the royal Dlamini clan, but includes members of other clans traditionally associated with royal power. Members of the ruling aristocracy are the governors of the royal homesteads, some chiefs, members of the Inner Council, prominent figures at the Swazi National Assembly, individuals such as the Great Councillor, and those royal family members who are appointees of the King in positions of authority within the bureaucracy.

Since the King controls and has power of veto over all appointments, individual power among the ruling aristocracy resides in the closeness of its association with the King. While its members may provide guidance, or act as advocates, personal policy making power is limited at the highest levels of government. The ruling aristocracy's enjoyment of royal patronage translates into personal power in its own domain, especially if the members of this category are chiefs. As representatives of thc King, the members of the ruling aristocracy perceive their role to be the perpetuation of royal power.

The economic power base of the ruling aristocracy resides in the King's direct control over the Tibiyo Taka Ngwane and Tisuka Taka Ngwane Corporations. These are the large parastatals which grew up after independence, and into which all mineral royalties are paid directly, to be held by the King in trust for the nation. Tibiyo Taka Ngwane was formed to invest this wealth for the benefit of the Swazi people, but the organization is not accountable to any branch of the government, profits do not accrue to the Treasury, and it is under the exclusive control of the monarch. Tibiyo Taka Ngwane was so successful that in 1975 it began to finance its own operations. Tisuka Taka Ngwane was then formed to manage the original Tibiyo Taka Ngwane fund (Booth, 1983, pp.106-107; Davies et.al., 1985, p.17). Booth describes the relationship between royal power and the corporations in the following terms:

> It appears to many, in short, that the Tibiyo Fund [Tibiyo Taka Ngwane] has become the vehicle for the item most conspicuously lacking to the royal house at the time of independence, a secure capital base for the reproduction of the monarchy (Booth, 1983, p.107).

The relative secrecy of Tibiyo Taka Ngwane operations is of concern to many observers. There is a lack of accountability, a number of joint ventures (involving large amounts of foreign capital) on resumed Swazi Nation Land, and commoners are unable to gain any real understanding of the operations of the scheme. Significantly, Tibiyo Taka Ngwane finances the Swazi Defense Force (Booth, 1983, p.107). Appointments to the higher offices of Tibiyo Taka Ngwane are made by the King, to whom the appointees are personally accountable. Tibiyo Taka Ngwane has thus become a resource base for the military and economic aspects of royal power.

Membership of the ruling aristocracy is determined predominantly, but not exclusively, by 'blood', and the rhetoric and social activity of this category are conservative traditionalism. The fixed point in the constellation of the ruling aristocracy is the figure of the King and that of his mother. Traditional linkages with the monarch facilitate entry to this category, but individual abilities remain significant.

The power struggle during the interregnum of Prince Mfanasibili was between prominent members of the ruling aristocracy, and their respective urban elite faction supporters. The urban elite stood to gain by a diminution of 'blood' as a factor of entry to ruling aristocracy power. Factional alliances achieved legitimacy during the interregnum only by supporting prominent royal family members. For example, Prince Mfanasibili, in his bid for power, could propagate uncertainty over the nomination of the child Nkhotfotjeni (later King Sobhuza II) as heir in 1899. Mfanasibili was the son of Prince

Makhosikhosi, brother of Sobhuza II and son of LaMavimbela. At the time of Sobhuza's selection as heir apparent there had been dispute and controversy, particularly because of the circumstances in which his and Makhosikhosi's father, Bhunu [Ngwane V], had died. The selection of heir was, as a result, carried out in an atmosphere of distrust (Kuper, 1978, p.31; Matsebula, 1972, p.9). The implications were that Mfanasibili's assumption of power incorporated elements of legitimacy because of the historical uncertainty associated with Sobhuza II's succession. Members of the urban elite who supported Mfanasibili were thus able to do so with a degree of rectitude.

Prince Mfanasibili had credibility as a prominent royal family member. He would not have been able to come so close to gaining complete power if he had not been a member of the royal family. In the end, ultimate political power could have been held only by a royal family member. While Swaziland retains the King as absolute monarch, the most prominent members of the ruling aristocracy will always be members of the royal Dlamini clan.

The urban elite

At the administrative level, the ruling aristocracy is strongly linked with another identifiable social category, the urban elite. While membership of the ruling aristocracy is predominantly determined by 'blood' and patronage, that of the urban elite is determined by education and patronage. Instigated during the colonial period among educated Swazi and white settlers, the urban elite burgeoned after independence, and now attracts young, educated people. The white settlers, once a separate class in colonial Swaziland, are part of this category and some have become prominent. For example, a member of a former settler family, Derek Von Wissell, is a cabinet minister, and Dr David Hynd, a missionary, was the Minister for Health for many years.

The former colonial indigenous middle class dominated the urban elite in the early Independence years. The few highly educated Swazi, men such as Arthur Khoza (*cf.* Khoza, 1979) and J.S.M. Matsebula, became prominent in colonial times and attracted the attention of the King. For example, Arthur Khoza received his first degree at the University of South Africa, then studied politics at the Ideological Institute of Winneba in Ghana. He was appointed to the civil service on his return to Swaziland and became involved in politics, eventually becoming secretary general of the Ngwane National Liberatory

Congress, from which he resigned in 1966. By 1968 he was secretary to the Prime Minister Prince Makhosini and, by 1973, secretary to the Inner Council (Kuper, 1978, pp.273, 298, 338). During the interregnum he resisted the Mfanasibili faction and fled to Zambia. Khoza returned to Swaziland in 1986, immediately after the coronation of King Mswati III, and is now a prominent and respected man.

J.S.M. Matsebula exemplifies members of the present day urban elite. A prominent historian and former secretary to King Sobhuza II, Matsebula is a member of a prominent 'ritual' clan,[1] and his mission education was funded anonymously by a British woman. He attended the new Swazi National School in 1935, and became known to the King. He qualified as a teacher in South Africa, and taught in Swaziland before enrolling as an external student at the University of South Africa in 1943. He taught in South Africa in the capacity of headmaster before being invited by the King to act as his *Lisolenkhosi* ('King's Eye' or liaison officer). This position was not unlike that of a private secretary. He became education officer, a position in the colonial administration catering for Swazi students. Later, Matsebula studied for a year at the University of London. In 1967, he became private secretary to the King, before retiring in 1974 after a long association with the King. On retirement, he became the chief executive officer of the Swaziland National Trust Commission, wrote extensively on Swazi affairs, and received a Doctorate of Letters from Swaziland University in 1978. He finally became Chairman of the SiSwati Development Board in 1980, a position he still holds (Matsebula, 1983).

In both these cases prominence was achieved initially through educational opportunities. Since Sobhuza II was also educated, he emphasized the need for educated bilingual Swazi advisers. As well, the colonial administration relied heavily on such people in the bureaucracy. Younger Swazi sought to emulate such men for they soon realized that education provided a key to progress.

The position of the urban elite is strengthened by the degree to which education is equated with knowledge and wisdom by the general public. In the past, limited social mobility forced commoners to trust in the wisdom of their leaders. Since educated Swazi assumed authority under the colonial government, and the King used such people, improved status acquired by education implied wisdom and knowledge, virtues rewarded with public trust. As such the urban elite provide powerful role models for young Swazi.

The urban elite exercise great control over bureaucratic processes, and are capable of dispensing significant patronage. Members have close links with the ruling aristocracy based on mutual need. They also have the greatest contact with external sources of capital. They decide the form investment programmes in Swaziland should take, and they control the issuing of licenses and permits for trade, or for business. Abuse of position is rife and some major scandals have erupted over excessive salary payments, the issuing of licences to South African companies seeking to avoid international sanctions against that country, misappropriation of public money, and even the acceptance of bribes from lower bureaucrats seeking promotion.

The urban elite increasingly have absentee status on Swazi Nation Land, and instead tend to make major investments on Individual Tenure Land, to the degree that farm ownership is a status symbol. Insinuations concerning misappropriation of public resources to these farms are common, but press disclosures of corruption only ever target those out of favour. As a result, corruption is rife but exposure depends on the tenor of royal favour. For example, one young potential member of the urban elite told me that, although he knew it was wrong, immediately he had power he would use it to get a farm. His rationale was that since the land could not last for ever, he must secure his future.

The urban elite are economically and politically independent of the rural homestead. Thier members are a product of primary urbanization who have benefited greatly from the development effort of the 1970s, and their power is enhanced by the process of secondary urbanization which has occurred since. They are committed to urban living and emphasize education for their children, favouring better quality or elite schools in Swaziland.[2] The urban elite are also highly conscious of status, and western consumer items figure prominently in their lifestyles. Throughout Africa, this category of people are known by the term '*waBenzi*' (Lit., 'of the [Mercedes] Benz') and in Swaziland they are conspicuous consumers.

Personal power for the members of the urban elite lies in their ability to dispense patronage. There is room for corruption in these patron client relationships, and the urban elite are often seen as avaricious. Other pressures facilitate corrupt practice among the urban elite. As patrons, members are also supposed to be able to dispense goods and resources because of the custom of begging favours. Since this is a significant traditional practice, they must be able to finance their status. To avoid corruption by not dispensing patronage and resources would be considered anti-social and inappropriate.

The urban elite can rise to great heights but, unlike the ruling aristocracy, 'blood' does not protect their members from failure. For example, the fall of the commoner Nhlabatsi, the former Minister for Education, was as spectacular as his rise. Inappropriately exercising the prerogatives of a member of the ruling aristocracy, he was perceived as arrogant, and lost public sympathy and support. The maintenance of power by the urban elite rests on its ability to remain largely anonymous, serving a succession of ministers, and theoretically remaining non-aligned with any particular faction.

Wage and salary earners

Wage and salary earners are a fast growing category, with a long history of wage labour in the South African mines, and in the colonial administrations. Swazi labour involvement in South Africa declined after the 1930s when local wage opportunities emerged in mining, education and agri-business. After independence, further labour was needed with the growth in urban areas and the establishment of light industry and hospitality industries in the towns. Most wage and salary earners are now employed inside the borders of Swaziland.

A clear relationship exists between the urban elite and the wage and salary earners, as the original urban elite grew out of this category. This category is best understood in two sub-categories; wage earners and salary earners. The sub-category of wage earner is then necessarily further divided into rural and urban based wage earners.

Salary earners As a sub-category of wage and salary earners, the salary earners are furthermost from the rural homestead and closest to the ideal of the urban elite. They are well established in the urban and peri-urban areas, and are usually not an integral part of the rural homestead, but link in with it at many levels. The point at which salary earners are distinguished from the urban elite is tenuous, because the difference relates to comparative power and the proximity of urban elite individuals to royal power. Women figure prominently among salary earners, because, like second sons, they are drawn to opportunities in the city. In this category I have also placed urban self employed people who have a primary commitment to acquiring their income from urban activity.

The salary earners have a multi-faceted involvement with the rural homestead, and are important patrons. They are constantly called on to provide supplementary finance for the rural homestead, and accommodation for those visiting or working in the city. Many children attending secondary school are dependent on these patrons, as boarding accommodation is relatively scarce and comparatively expensive in the towns. For example, one prominent Real Farmer sent all his children to a good urban secondary school because his brother was prepared to accommodate them. This has developed into a long term arrangement whereby the family always stays with the brother when in town. In this particular family, the grandmother coordinates a network of familial patrons and clients, whereby a great circulation of children and resources is maintained. Placing a strong emphasis on family stability, she arranges for her grandchildren to be looked after in stable homes and homesteads. For example, while one uncle provides accommodation for teenage children in the city, the Real Farmer provides accommodation and education for an array of children on the country homestead.

The economics of this procedure are most effective. Good secondary schools are located in the urban areas, and are expensive, prohibitively so if accommodation is included. Primary schools in the rural areas are readily accessible and inexpensive,[3] and the rural homestead is fundamentally self sufficient in food. Large numbers of children in the country are easily looked after, and provide valuable labour. Young salary earners can thus work in the city and send their children to the country, if they are prepared to accommodate teenagers who need secondary education.[4]

The urban patron also acts on behalf of the wage earner category, and specifically for the rural based wage earners, by providing contacts for prospective job opportunities in urban areas. The labour market is filled on an informal basis, and rural people find it difficult to get work without an urban patron. The patrons monitor the market and notify the rural homestead of potential opportunities.

Salary earners also provide a stable environment for those moving to the city, since accommodation in urban areas is scarce. The urban patrons provide a safety net in the transitional stages of migration, particularly as first wages and salaries often take considerable time to come through. Some people I know waited three, four and even five months before receiving their first pay, an impossible situation if they were not supported by an urban patron. For example, one young man successfully completed higher secondary schooling, but was unable to get a job. He worked for his father at Ntimane, and eventually won a paying apprenticeship in town. He moved to town and lived with a female relative of his family. She owned a house and her domestic situation was stable. Immediately he encountered both logistical difficulties and problems with subsistence financing. His employers, a government department, failed to pay him for approximately four months, and made no attempt to cover his subsistence needs during that time. Despite the inconvenience, he managed well and none of his family expressed concern at his situation. The urban patron subsidized him until his wages came, and his costs were offset, to a degree, by remittances of produce from the country.

While the relationship between the rural homestead and the urban patron can be mutually rewarding, it is not necessarily symbiotic. Some urban patrons complain that rural kin take them for granted, and overestimate their capacity to support their rural relatives on extended visits to the city. As the daughter of one urban patron complained to me: '... they [rural relatives] think we have so much money that they don't even bother bringing maize anymore'. Despite this, however, certain benefits do accrue to the urban patron. For example, political identity, for those without reasonable private land holdings, is still centred primarily on Swazi Nation Land. Political representation is still usually made through the rural homestead, since the Regional Assemblies are rural based, and access to the chiefs and many members of parliament occurs most effectively through the homestead headman. Also, urban males stand to lose the right to give allegiance, and so must maintain essential links with the chiefdom of their patrilineage on their sons' behalf. Women enjoy obvious benefits in being urban patrons. Limited by gender to secondary roles on Swazi Nation Land, and awarded minority status on marriage, female patrons are able to pursue careers and accumulate property, yet still use the rural homestead as both a creche and a resource base. Many women are thus able to avoid marriage altogether.

Salary earners are not all resident in urban areas. Teachers and nurses in rural centres are part of this category, as are the bureaucrats associated with the various development organizations. Many of the large agri-business industries have built fully serviced townships in the rural areas for upper management Swazi and expatriate salary earners. Salary earners in this position find it difficult to act as patrons, since their ability to accommodate people is circumscribed by management. Thus, many rapidly become alienated from the rural homestead, and the socialization and standards of education of their children differ from those of other Swazi children. Salary earners in these circumstances are closer to the urban elite ideal, but socially and culturally more isolated.

Salary earners are part of both planned and unplanned migration to the city. For many, lack of opportunity on Swazi Nation Land has resulted in a planned move to the city. Eventually, as urban patrons, they provide the means by which the rural homestead is consolidated on Swazi Nation Land, in a two way flow of benefits. Other salary earners find themselves in the town by default, and once there, take opportunities which arise. In time their status as urban patrons may acquire all the appearances of planned migration, despite the unplanned nature of its origin.

Wage earners The category of wage earners is further divided into the sub-categories of urban wage earners and rural wage earners.

Urban wage earners can either be resident in the urban areas or, alternatively, can reside in rural areas but remain primarily reliant on urban wages. The point of entry to urban wage earner status rests on personal commitment to the rural homestead (see Chapter Three). The uncommitted rural dweller easily begins to lose contact with the rural homestead when involved in urban wage work. Many wage workers are absorbed into the compounds of large agri-business companies, and contact with the homestead becomes sporadic.

In Ngwempisi, many young men enter this category, especially when they come from marginal homesteads. Unlike large homesteads with kin groups extending into the towns, and a constant two way flow of resources and social contact, the marginal homesteads have little to offer. As they become reliant on remittances, an essential balance of authority is lost, and control lapses.

People become urban wage earners through both planned and unplanned migration. Many young people go to the city looking for work as part of a planned strategy and, on moving away from the urban patron, gradually lose significant contact with the homestead. Other young people drift to the city in an unplanned manner, quickly become dependent on wage earning, and often have little kin based support as a safety net. In this category I include the urban unemployed who can move in and out of wage work.

In this sub-category, social control is less tangible, causing some concern for the authorities. For example, the notorious *sidlanis* (hoodlums), who are blamed for much of the disruptive activity on the streets at night, are either members of this group, or are the children of this group. Since unplanned urban migration usually involves relatively uneducated males and some females, opportunities for advancement are limited, and pronounced social problems are evident. The infamous *Ekhaya* (go home) programme of the former Prime Minister Bhekimpi was directed primarily at the people in this group. The Ekhaya programme targeted the occupants of the shanty towns on the fringes of Mbabane and Manzini. The Prime Minister envisaged that the unemployed would be sent to the country to till the soil. The programme was a notorious failure, because the public perceived it to be an obvious attempt by the Prime Minister to use the programme to divert attention from the troubles of his government.

Rural wage earners are the sub category of people who work in urban and rural wage earning capacities, but do so as the result of a planned strategy, sustaining a high degree of commitment to the rural homestead. Their work environment is similar in all respects to that of urban wage earners, but unlike the latter, they stay within the kin group and, if they become unemployed, they return to the homestead, or are supported until they find more work. The men in this category are usually either rural headmen, the heirs of rural headmen, or men intending to become rural headmen. The women are usually mothers in a rural homestead, or the women of a homestead on the urban fringe. Like the urban wage earners, the rural wage earners work in low paid manual jobs.

Rural wage earners are attracted to the low paid manual work by the flexibility of employment, rather than solely the need for cash. The advantage of wage work is that little time is spent building an urban social profile, thus allowing the maintenance of a rural profile. This is impossible in salaried work, especially when committed headmen, or intending headmen, are under even greater pressure to enhance their rural social profiles to gain advantages in a competitive environment. For example, one talented and educated young man had to choose between a salaried city job and his desire to become a commercial farmer. He recognized his own potential to achieve urban elite status in the city, and Real Farmer status in the country. This presented him with a considerable dilemma, for any time invested in realizing his potential in one area necessarily reduced his effectiveness in the other.

The young man solved his dilemma by building profiles in both areas simultaneously. He was offered a teaching position in an urban school, a choice only given the best graduates. He then approached the headmaster of a school near Ngwempisi, and contrived to have the headmaster request the Education Department to send him to that school. This course of action allowed him to continue living and socializing in Ngwempisi, but ensured he was a member of the public service. Though, in this instance, the young man earns a salary, the choices he made resemble the choices many wage earners make, which allow them to invest as much time as possible in their rural social environment.

Many wage earners work to finance a rural homestead. For example, one Real Farmer who is not a member of Ntimane, cultivates an irrigated garden near the Mtimane. He had given allegiance to the Chief in 1974 but, as a second son, received little land. He then became a rural wage earner while cultivating vegetables at his homestead. In 1982 he became a full time commercial vegetable producer. He is now a significant Real Farmer, and is sought after as a potential member of Ntimane. Wage work allowed him to enter cautiously into commercial production, while building a significant public profile in Ngwempisi. He left his transient wage labourer status far behind him and currently, despite having none of the opportunities of first sons, is becoming a significant person in the chiefdom.

For many established headmen, wage labour gives them the vital flexibility they need to maintain their rural political and social profile. These headmen, who are common in Ngwempisi, return every weekend to their homesteads. An illustration of this a headman whose wife is a member of Ntimane. A well respected man of great dignity, he is a rural wage earner, has a significant political profile in Ngwempisi, and is considered an expert on traditional matters. He is consulted extensively on matters of etiquette, dress and display. Though not a Real Farmer like his wife, he is a Real Person, and is present at all important occasions. He has achieved prominence through the force of his personality, but is only prominent among those in Ngwempisi, and through constant social activity. Wage labour serves this man's purposes well, and a salaried job or an urban commitment would negate the richness and diversity of his life in the country. An added element for him is his wife's membership of Ntimane, and the benefits that accrue from this situation. In the long term, as he gets older, there will be less need for him to work and he can retire to the country to be supported by the Ntimane garden. In the short term, his wage work has paid for agricultural inputs, education for his children, and consumer commodities.

This use of wage labour is most clearly seen among older established headmen, often representing conservative and powerful factions on Swazi Nation Land, who simply need to maintain their homesteads. There is no correlation between their occupational status and their status within the chiefdom. For example, the brother of an important Real Farmer lives in a kin cluster of homesteads which are among the most prominent in the chiefdom. He is a notable member of the Chief's councils, and is a leader in various civic projects. For example, he coordinated the funding and building of a secondary school in the area, using a E10,000 donation from the Canadian government. Yet despite his brother's prominence in commercial farming, he has not aspired to commercial farming, relying instead on his own wage labour, initially in South African mines in the 1940s, and then in Swaziland. He is unquestionably a member of the Real People, and supports Real Farmer initiatives before the Chief.

Wage earners in this category may maintain an urban dwelling, but it is always an extension of the rural homestead. All wages are remitted to the rural homestead and as much time as possible is spent in rural areas. It is from among such people that commercial farmers emerge. Even limited commercial opportunities in rural areas are an attractive option to urban work and its potential for alienation.

Another prominent Real Farmer worked for Usutu Pulp, then moved to work in a shop in Mliba so that he would be even closer to his home. During that time he never lost sight of his primary commitment to rural living. Inspired by his Real Farmer father-in-law, he became a commercial farmer in 1970. The transition was marked by caution. He told me that he had a variety of reasons for wanting to join Ntimane:

> I decided to join Ntimane as I wanted my wife to work on the garden. I was still working in Mliba at the time. I saw that working for someone else didn't give me enough income. They [employers] didn't even allow me to be five minutes late. If I was late they deducted it from my wages. It was like working with oxen; you [use them to] plough the land and when you are

> finished you don't give them the crop. When I was working I didn't even have time for a break to go to the toilet. When I asked for a raise they didn't increase it [the wage].

This man and his wife achieved commercial viability at Ntimane by coordinating the man's wage earning efforts with the woman's efforts at the gardens. Subsequently they were able to fund their children's education entirely through the proceeds of their farming enterprise. Both the man and his wife told me that they made a good decision, earning more than they could have in wage employment, and being able to offset their costs by initial wage earning. Significantly, their children are planning similar careers, and express strong commitments to farming. Nevertheless, urban wage earning is still an important means of acquiring capital to commence these ventures.

In general, wage earners are a major transitional group, bridging the rural/urban divide. When considered in the context of commitment to either a rural or urban lifestyle, wage earners are categories in considerable flux. Unemployment and self employment are alternative states for those within this category as people react to opportunities that arise, or are not successful in their endeavours. Urban wage earners become increasingly marginalized once the kinship linkages with the rural homestead are broken. Rural wage earners retain their rural linkages and contemplate their return to full time living in rural areas. In both cases, eventual success can only be obtained by a full blown commitment to either urban living or to rural living. Urban wage earners hope to become salary earners, and rural wage earners look to their future on Swazi Nation Land.

Rural commoners

Rural commoners are the social category which conforms most closely to that of the traditional model of Swazi commoners. Rural commoner economy is essentially subsistence, and highly dependent on maize production and animal husbandry. Rural commoners are reliant, to varying degrees, on remittances and limited commercial activity on Swazi Nation Land. They are not completely self sufficient, nor can they be. They are distinguished by their commitment to the rural homestead, and are the base unit from which come the rural wage earners, the urban wage earners, the salary earners and the Real People. The rural commoners now tend to be older, increasingly marginalized people.

For example, KaGwebu is the homestead of a headman, his wife and children. The homestead is relatively poorly sited, and has limited fields. Nevertheless, the homestead economy is sustained by cultivating and some sporadic wage earning by the headman. The children are young, and until such time as they leave to work in the towns, cash flow into the homestead will continue to be limited. Gwebu is not a prominent headman, and he does not have the potential to become one. He does not have strong contacts in the chiefdom, and nor does his wife. Significantly, quite recently, the headman's wife LaTwala has struck up a friendship with the wife of the aging owner of plot fifteen, and has been helping on that plot in return for vegetables. Over the long term this could be very significant, because if she proves herself, it may grow into an opportunity to gain entry to Ntimane, and vastly improve the fortunes of the family at KaGwebu.

The Real People

The Real People are the category of people described in Chapter One. This group is immediately distinguished by its commitment to living on Swazi Nation Land. It comprises a number of sub categories: Local government personnel, the Chief and the headmen, and the Real Farmers.

Local government personnel These can be distinguished from urban salary earners, because they are resident in rural areas. Local government personnel can only become Real People under certain circumstances specific to personal interest and commitment. Some have gained good reputations among rural dwellers, and as a result, have acquired significant profiles. Some are also valuable to local farmers for the range of contacts which they offer. Those who are trusted provide vital insights for farmers into the nature of the workings of the government bureaucracy, and facilitate a reconciliation between the often arbitrary will of the central bureaucrats and the rural people. However, as they are essentially urban oriented, local government personnel may easily come into conflict with Real Farmers, the Chief and the headmen, all of whom have a primary commitment to rural living. The conflict grows out of the urban focus of the central government which tends to be at odds with the rural focus of the Real People. The ability to resolve such conflict is often the point of entry to Real Person status for these personnel, and distinguishes some from the others who remain rural based salary earners.

For example, while an array of programme officials are resident at the Ngwempisi Rural Development Area headquarters, only one, the frontline extension worker, is considered a Real Person. His status rests entirely on his attitude to rural people. The other personnel maintain a distance and are aloof from the local people, avoiding social contact except on formal occasions. Possibly the youthfulness and urban backgrounds of these functionaries make it difficult for them to appreciate social dynamics in the rural areas. Furthermore, their tendency to advocate egalitarianism as appropriate in giving access to Swazi Nation Land, and for entry to, and the governing of, the irrigation schemes, brings them into conflict with the Real Farmers, the Chief and conservative headmen.

The aloof manner of these personnel prejudices them in the eyes of the local people, because, as I described in the case of the drunken official, such a demeanour is associated only with those in positions of power and authority. The farmers quickly come to resent their indifference when, unlike traditional authorities, these personnel are unable to deliver on their promises.

The new pipeline from the dam on a nearby stream to Ntimane was a good illustration of this situation. The government personnel were indifferent to the difficulties, the strain and the financial burden they imposed by their inefficiency over the five year period. Initially planned as a three month project by a Scottish engineer, it lapsed under the control of his Swazi replacement, who left the implementation of the project to an untrained labourer. The three month project extended to five years as the farmers were directed to build pylons and lay pipe, only to be told to dismantle them. Timber, cement and other components were wasted in large quantities, necessitating re-purchasing and additional cartage. Supervision was sporadic and machinery required for specific tasks arrived irregularly.

Despite their irritation, the farmers persisted and eventually, under the guidance of a new engineer, succeeded in gaining their water supply. Although angered by their treatment at the hands of the previous engineer, they are impressed with his successor. Throughout this period the local frontline extension worker was often present and constantly expressed sympathy for the farmers' plight. The farmers enjoy their contact with both this officer and the replacement engineer and consider them both reliable and likeable.

Members of the local government personnel are also seen as intrusive when they try to involve themselves in marketing. Farmers doubt the efficiency of centrally controlled markets, and prefer their own individualized marketing strategies for the control they exercise over quality and sales. The farmers appreciate their own vulnerability in any relationship with the bureaucracies, and a crisis emerges often enough to drive the point home. For example, the 'maize crisis' in 1985, during which the only mill in Swaziland was closed, discouraged many farmers who had considered growing maize commercially. In 1985, because of government interference, the only commercial maize mill in Swaziland was taken over by the government and immediately became inoperable because of high overheads. Simultaneously, licences were given for maize to be imported from South Africa to make up the deficiency. Storage facilities at Rural Development Areas Programme centres, built at huge cost, were still not finished in 1989, so farmers can neither store nor sell their maize.

The transition of some local government personnel to Real Person status is dependent entirely on their interest in regional issues. Now that the Rural Development Areas Programme has ceased growing, it will possibly stabilize around more motivated and expert personnel who recognize the significance of the Real People and the Real Farmers. That Real Farmers propose having their own educated children return to the area as experts is significant, since it suggests that functionaries who wish to be Real People will need to accept a situation where much more control is exerted over them by their rural peers.

The Chief and the headmen Within the social category of Real People, the Chief and those who surround him are especially significant. In all circumstances, the Chief relies for the infrastructure of his administration on the headmen, who are all those who have given allegiance in his area. The headmen are the Chief's power base, whether permanently resident in the area or not. The headmen are thus part of many categories, though traditionally they were predominantly rural commoners. They now form part of a range of categories committed primarily to rural living. Factionalism is evident among headmen when some, particularly in the rural wage earner category, ally themselves with rural commoners, to form a conservative power bloc in opposition to Real Farmer influence. Generally though, prominent headmen, as Real People, tend to support Real Farmer initiatives.

The focus of division within the headmen category is commercial farming. Those who do not intend to produce commercially on Swazi Nation Land are in basic opposition to headmen with commercial goals. Increasingly, the Chief is gathering headmen around him who conform to Real Farmer standards. As general acceptance of this situation develops, the headmen who do not conform are identifiable as a sub-category of aging, or marginalized, headmen.

The final sub-category within the Real Person category is that of the Real Farmers, whom I have examined in detail in Chapter One.

Swazi social categories are imprecise and transitory, though clearly emerging from pre-existing social formations. The imprecision of definition is a result of the necessity for most individuals to enter into commercial activity. The constant movement of people into the towns, and back to the country, makes difficult traditional methods of enforcing conformity. Furthermore, material wealth is another factor of differentiation which overrides traditional status in some circumstances. The categories and sub-categories tend not to incorporate anomalies such as unemployment and self employment in urban areas, though clearly these are areas in need of further research. The activities of the Real Farmers, and their emergence as a recognizable sub-category, are both a stimulant for, and a result of, intensity of change in the social environment of Ngwempisi.

Notes

1. The Matsebula were valued for their ritual contributions to the Swazi royal house. The Matsebula are considered ritual 'in-laws' to the king as the first royal wife is always a Matsebula.
2. Schools like Waterford Kamhlaba, part of the elite United World Colleges system, provide an education for the children of prominent Swazi, expatriates, and overseas children (Zeigler, 1985, pp.663-666).
3. Primary school places are in excess of the demand among school age children (Government of Swaziland, 1984).
4. There are 468 primary schools in Swaziland but only forty-seven junior secondary schools and thirty-nine high schools. This means that only

seventeen per cent of eligible children leaving primary school gain entry to junior secondary schools. This is further reduced by the high number of 'out of age' students in secondary education. As a result, secondary school places are in high demand (Government of Swaziland, 1984).

8 Real Farmers, society and the state

Introduction

The Real Farmers uphold traditional authority and the norms of agnation, while, in the long term, facilitating the emergence of a cleavage between themselves as property owners and the rest of the general population in Ngwempisi. In this chapter I discuss this process in a broader conceptual context, demonstrating the linkages between the economic, social and political factors of Real Farmer influence. I then return to the issue of the role of the Swazi state in the emergence of the Real Farmers, and I show that Real Farmer activity is congruent with the re-emergence of Swazi royal family control over the state.

Custom and change in Ngwempisi

In his book 'Palms, Wine, and Witnesses', Parkin (1972) describes economic, political and cultural change among the Giriama of Kenya. Parkin noted that custom was both threatened and strengthened by the activities of Giriama entrepreneurs who were making long term capital investments in land rather than in people (1972, p.5). Parkin observed that the emergence of minor family dynasties, based on capitalist enterprise, involved the reconciliation of 'paradoxes of custom'. Parkin describes this in the following terms:

> In Giriama I can identify three such fields. Each proclaims an ideal which supports the re-distributional economy and gerontocracy.

> Yet, under the protection, so to speak, of this ideal, a new, more fundamental but initially illicit capitalistic principle is smuggled in. I call these types of mystification, paradoxes of custom (1972, p.99).

In Swaziland, the Real Farmers also reconcile paradoxes of custom by supporting traditional rights of usufruct, while seeking to gain proprietorial rights over land. The Real Farmers denigrate kinship as a factor in gaining access to community resources, but use kinship to exercise significant political control within Ngwempisi, and at Ntimane.

In Chapter Two, I argue that the Real Farmers emphasize the resolution of issues through consensus. This consensus politics has two levels. First, at a 'democratic' level, issues pass through the committee and are ratified by the members without dissent. Second, at a 'paternalistic' level, the Real Farmers use the customary deference Swazi have for their leaders to forestall dissent. Thus, consensus is achieved and solidarity maintained.

To achieve consensus, prominent Real Farmers play on public awareness of customary practice to achieve their particular goals. I gave the example of the Ntimane member who left the scheme after the Real Farmers used a witchcraft accusation to exert pressure on him. Throughout the process they avoided a direct, and potentially divisive, confrontation. The key element in the process was the moral rectitude that the Real Farmers communicated to the members because of the seriousness of the witchcraft charge. This rectitude is at odds with their stance on other members who are known wizards, but still practise fully as members of Ntimane. The Real Farmers focused their concern on the fact that the man had not cultivated his garden, which constituted an affront to the corporate work ethic of the Ntimane members, rather than to his alleged status as a wizard.

The Real Farmers achieve their goals through consensus, by manipulating public preferences for political paternalism. This form of political action has strong antecedents because of the active way in which King Sobhuza II promoted paternalism as an appropriate cultural reality. For the Real Farmers, the King's involvement in politics throughout the colonial period, and in the post independence era, was a clear demonstration of how to reconcile modernity and tradition. By taking the committee structure of Ntimane and using the committee to process non-controversial issues, or ratify the decisions of prominent Real Farmers on controversial issues, the

Real Farmers avoid letting democratic principles interfere with their expertise in customary procedures. The Real Farmers play on the Swazi fear that divisiveness leads to jealousy and witchcraft. While some younger men express reservations over the paternalism of the prominent Real Farmers, they accept it as fundamentally appropriate in the circumstances.

In Chapter Three I described how the Real Farmers have reacted to looming land scarcity by trying to gain proprietorial rights over their homestead land. For the Real Farmers, the process of gaining proprietorial control over land involves resolving a series of conflicts of interest at different levels in the chiefdom. I described how the Chief's status has traditionally been enhanced by the number of retainers he had. As a result, in trying to acquire land, the Real Farmers are potentially in conflict with the Chief's desire to accommodate more retainers.

The Real Farmers respond to this situation by supporting the Chief's right to allocate the remaining marginal land to young men who have given him their allegiance, but by influencing the choice of candidates eligible to receive land. The Real Farmers, through the involvement of some of their number in the Chief's councils, support the candidates who are, like themselves, committed to farming in Ngwempisi. In many cases these candidates were not born in the area, and are not of significant lineages. Therefore, while the Real Farmers support the traditional process of giving allegiance in return for land, they also downgrade the importance of lineage and locality as integral to successfully getting land in Ngwempisi.

The Real Farmers are involved in significant non-customary behaviour at their homesteads. That is, the Real Farmers, using land through traditional rights of usufruct, are seeking proprietorial control over that land. I described this process as one involving the deliberate and systematic under use of homestead land through a variety of techniques. These techniques include using traditional farming methods, restricting available labour, the restricted use of land, and the use of fencelines and natural features to approximate boundaries.

The Real Farmers are only able to subsidize their under use of homestead land by utilizing the resources of their irrigated gardens at Ntimane. The complementary trend among Real Farmers to monogamous marriage united the formerly divergent interests of the wife and the husband in a polygynous marriage. I argue that the conjugal family is based on a strong imperative; to provide for the children in a future in which the Chief has no land to give.

Therefore, the acquisition of proprietorial control over land by conjugal families results in sons focusing their loyalties on their parents, rather than on the Chief. Similarly, this also involves family members rejecting the customary belief that land given under rights of usufruct could be reclaimed and re-allocated by the Chief according to need. Instead, by claiming land and determining to pass that land to their children in the future, the Real Farmers ultimately contradict the Chief's traditional control over land. Yet it is these same Real Farmers who act on behalf of the Chief in resolving sensitive issues pertaining to land, through their duties as members of the Land Management Committees. Once again the Real Farmers manage to publicly uphold customary practice in the short term, while striving to achieve contradictory long term goals.

Significantly, the Real Farmers, through their actions, are making land inaccessible to many other people. Rather than supporting the Chief's customary practice of accommodating as many retainers as possible, the Real Farmers are restricting his ability to do so by giving rights to use of land a particular moral value, a value at odds with rights of usufruct.

In Chapter Four I described how the Real Farmers support certain customary practices, because by doing so they can further under use their homestead land. The Real Farmers differentiate between the labour they use at Ntimane, and that which they use at their homesteads. Homestead labour patterns emphasize kinship, and the labour used is drawn from among kin. In the examples given, I showed how kin based labour allows the Real Farmers to build significant profiles among their kin. In contrast, the Real Farmers restrict their use of labour at Ntimane to their immediate family, or to some wage labourers, usually Mozambiquan refugees, whose efforts are denigrated.

By differentiating labour in this fashion the Real Farmers seem to discriminate between their traditional use of land at their homesteads and their commercial activities at Ntimane. Thus the Real Farmers avoid antagonizing their neighbours and relatives who are not members of Ntimane, by refraining from developing patron client relationships with them. If they allowed such relationships to emerge quickly, they would inevitably cause jealousy, with witchcraft and social unrest as a predictable result.

Another area in which the Real Farmers knowingly contradict customary practice is that of women's work. I described how male Real Farmers have adopted tasks formerly considered the duty of women, and how the Real

Farmers, male and female, have attributed to this work positive values which they describe as christian values of honesty and industry. The result of women's work acquiring non-customary status has been a rapid change in the status of women, a change the male Real Farmers support and endorse. In this case the Real Farmers, while supporting the traditional and exclusive right of males to land, are actually facilitating the circumstances whereby women are gaining rights to land to the exclusion of some men. This is a radical departure from customary practice. Not only are Real Farmer women in Ntimane in control of land, but, as I have described, there are circumstances where Real Farmers are influencing the allocation of land, through women, in the general community.

The Real Farmers support the exclusive right of males to land through the act of giving allegiance. They also facilitate women's access to land through conjugality and membership of Ntimane. Thus, they are devaluing gender as one of the major customary determinants of eligibility to gain access to land, and are replacing it with a higher value on labour and farming ability.

In the area of market transactions the Real Farmers also manage to be perceived as supporting customary practices while, in fact, achieving particular non-customary goals. I described how the Real Farmers refer to their sales at Ntimane as commercial sales, and to their clients as customers. While they use these impersonal terms, they engage in the substantial redistribution of vegetables by selling to local customers at low prices. At times they even give away large quantities of vegetables. In the short term, the Real Farmers are seen to support the customary practice of begging. Customers are able to exert considerable pressure on farmers to give them vegetables. Because of the custom of begging, the customers do not lose status through this activity, and are not seen by others as clients of that particular Real Farmer. In the longer term, the Real Farmers protect their enterprise by giving vegetables away, because vegetables are sufficiently plentiful that most local people will not bother to grow their own. In this way the Real Farmers protect their dominance over commercial activity in the area, and are able to continue subsidizing their under use of homestead land.

The Real Farmers also support the traditional right of people to graze their cattle on Swazi Nation Land. This is a fundamental right of all Swazi, and can be equated with the right of men to get land. The Real Farmers have not tried to commercialize their cattle herds. In fact, as I described in Chapter Four, they continue to own cattle which are as indifferent as those of the general community. Because they support the general right of all Swazi to graze cattle, the Real Farmers are seen by the community as preserving the customary significance of marriage cattle. In fact, the Real Farmers use their cattle as collateral to get loans for Ntimane. The Real Farmers avoid 'loan indiscipline' (see Chapter Four) and thus benefit from the decision made by the major Swazi lending institution to only lend money on cattle collateral to corporate bodies such as Ntimane. In this way, by supporting the customary right of cattle ownership, the Real Farmers use the estate of the general community to ensure their continued access to finance.

In Chapter Five I described how the Real Farmers support the notion of Ngwempisi as a cohesive community, with the Chief at the apex. They do this by de-emphasizing the difference between themselves and the members of the general community, by displaying obvious respect for the Chief, and by supporting community projects. In reality, the Real Farmers at Ntimane are, as I have discussed, increasingly differentiated from the general community. Within Ntimane can be found differentiated control over land, a strongly articulated set of community values, an economic support base, and a internal, legitimate, judicial authority.

In the general community the Real Farmers, despite their attempts to appear inconspicuous, exhibit some differences derived from Ntimane. Their relative affluence shows in many small aspects of material culture. Nevertheless, they work hard to reassure their neighbours and kin that they consider the general community of paramount importance. This is most obvious when an individual Real Farmer holds a work party at his or her homestead. Ambitious Real Farmers use this institution to engage in legitimate displays of wealth without offending against custom. Once again, the Real Farmers uphold recognized social institutions such as work parties, and earn the esteem of their relatives and neighbours. In the long term, the exercise enhances the influence of the Real Farmers because, as time passes, others who are not Real Farmers cannot marshal the resources to compete through such displays.

In Chapter Six I described how the Real Farmers are coming to share the authority and power of the Chief. Once again it is clear that the Real Farmers support the traditional authority of the Chief, but in the long term gain significant authority and power themselves. Particularly noteworthy in this process is the manner in which the Real Farmers act in support of the Chief against the acquisition of power by local government officials. The Chief's power has moved away from his ability to dispense land toward a situation whereby he legitimates, through his approval, the actions of his retainers. The Real Farmers are careful to seek and retain the Chief's support, and in return have won his backing. The relationship between the Chief and the Real Farmers is currently symbiotic as the Chief is enhanced by the Real Farmers, and the Real Farmers are enhanced by the Chief. Significantly, the Real Farmers benefit from the Chief's ability to control the ferment of witchcraft, directed against those who incur the jealousy of their fellows.

The Real Farmers are particularly advantaged by the Chief's control over the various organs of government in Ngwempisi. The Real Farmers know that while they maintain the Chief's support, they need not fear interference from outside Ngwempisi. However, the Chief's support lasts as long as he lives. The Real Farmers acted to forestall a loss of support prior to the Chief's death. For example, as I noted, the mother of the Chief's heir is a member of Ntimane. In other ways the Real Farmers have cultivated alliances with some local government officials, making it possible for them to join the ranks of Real People in rural areas. The Real Farmers are especially interested in having their own children return to the area as educated and influential government officials. The Real Farmers recognize that the Chief's power is diminishing. They support him, but they also use him to achieve their goals. While they do this, promoting tribute labour and seeking his judicial backing, they also increasingly share his authority. In supporting him they operate in his interests for the moment, but ultimately they act in their own interests and those of their children.

In Chapter Seven, I discussed the form social differentiation has taken. Once again people within these categories, like the Real Farmers, are seen to support tradition while striving for particular goals. With the exception of the ruling aristocracy, the need to forthrightly reconcile custom and change is lessened the closer one is to an ideal type of role. Therefore, the urban elite conform strongly to the model of urbanized educated Swazi, who are confident in their position and have little need to proclaim customary ideals. At the other extreme are the rural commoners, who also conform closely to an ideal model of traditional Swazi, and so do not need to constantly seek reconciliations of custom and change.

The ruling aristocracy and the King have promoted, in the public mind, a faith in paternalistic government by the King. By constantly reiterating the traditional and ritual significance of the King to the well being of the Swazi people, the ruling aristocracy and King have successfully reconstituted their pre-colonial dominance over the Swazi. As I described in Chapter Seven, they have done this through developing a significant economic power base through the royal family companies. The ruling aristocracy continues to actively propagate the notion of Swazi society as highly traditional by placing considerable importance on ceremonies and customs. Despite this, the children of the ruling aristocracy enjoy elite educational opportunities and are highly urbanized, though prominent in their traditional dress on significant ritual occasions.

There are obvious advantages for the ruling aristocracy in supporting their traditional status, and they are able to avoid public non-customary behaviour by delegating such work to their urban elite representatives. As a result, the ruling aristocracy is only seen in public as traditional figures. Members of the urban elite are not active proponents of customary practice and are not expected to make public statements in support of custom, leaving such activities to their ruling aristocratic patrons. The most striking contrast between the ruling aristocracy and the urban elite is in their public appearance, with the latter favouring western clothes over traditional clothes.

Wage and salary earners encompass almost the entire range, from the urbanized salary earners who closely resemble the urban elite, to the rural wage earners who most closely resemble rural commoners. The salary earners, like the urban elite, are distant from the rural homestead and experience less ambiguity in reconciling tradition and change. Many salary earners are also urban patrons, and as such they provide critical linkages with the rural homestead, and continue to support the rural ideal. I have suggested that there is an economic and political motivation in acting in this manner. Urban patrons retain linkages that allow them to redevelop rural profiles, and they often use educational opportunities in the countryside for their primary school age children.

The urban wage earners tend to conform to a model of peri-urban wage earners in their tenuous ties to their rural homesteads. As such, they have limited need to support traditional ideals to achieve their goals, and as a result social unrest, usually aimed at employers or political leaders, is most evident among this category. Members of this category are also the source of greatest concern to the ruling aristocracy and the urban elite, because of the limited degree of control attainable over them by reference to custom.

The rural wage earners, on the other hand, constantly adjust their perceptions of customary practice to achieve their goals. They are in the difficult position of having to convince their rural chief that they are committed to rural living, and to convince their urban employer that they are sufficiently committed to hold a job. Because rural wage earners ultimately intend to live on Swazi Nation Land, they tend to secure wage work which allows them some flexibility. Rural wage earners strongly support their traditional rights to land, but also consider change advantageous, in the manner of the Real Farmers.

The rural commoners, like the ruling aristicracy, ostensibly conform so closely to an ideal model, that they seem to experience few ambiguities in reconciling custom and change. However, some rural commoners adopt highly tradition orientated lifestyles, while others seek opportunities to move out of that category. The latter form the basis of all other categories, while the former, at times, react strongly against change. This reaction often appears to be an attempt to cope with the rapidity of change about them. By retreating into a dogmatic advocacy of tradition, without the ability to reconcile customary practices with the practices of the Real Farmers, these individuals are becoming increasingly marginalized in their traditional social environment.

The Real People were discussed in detail in the first part of Chapter Seven. However, the local government personnel sub-category I have identified does involve individuals who, to varying degrees, are in a process of supporting tradition while achieving their own particular goals. I concluded that those personnel who are not committed to rural living, nevertheless, do support the existing power structure and the values articulated, with the long term goal of becoming an integral part of the bureaucracy. The functionaries who seek stronger links with the rural community accept the dominance of values expressed by their rural peers. To succeed, they must also support the institutions that other Real People support, and they must ultimately accept the long term goals of the Real Farmers.

The activities of the Real Farmers in upholding tradition constitute a concession on the part of individuals who are challenging the existing order. These actions do not necessarily constitute a premeditated strategy on the part of Real Farmers. In fact, in upholding tradition, most Real Farmers say that they believe implicitly in the integrity of custom. For the observer, difficulties arise in ascertaining the values implicit in the decisions Real Farmers make, and in choices made by the members of the general community, and the base level at which these values are orientated towards, or away from, the community, however defined (cf. Epstein, 1962, 1973; Foster, 1965, 1973; Hill, 1986; Sahlins, 1958, 1961, 1969, 1972).

The primary motivations of the Real Farmers, in their coordinated attempts to bring about change, do not necessarily reflect a commonality of purpose. Therefore individual Real Farmers may have markedly different long term goals. Some may have goals which ultimately support existing community structures, while others may see such gaols as representing a transitional stage to another longer term arrangement. For example, in conversation with two leading Real Farmers, I discovered that one supported the cooperative organizations because, as he said:

> I like the way people work together. When I am in Ntimane I feel that I can get much done. There are many people to help and we are strong this way. No, I don't want to be a farmer alone without the cooperative.

On the other hand, the second Real Farmer had a very different long term goal. He was planning to use the money he made through farming at Ntimane, and that which his son made through wage earning in the city, to purchase land under individual title. He argued that he preferred this because:

> I believe that I am gifted in raising animals and now I raise vegetables because I don't have enough land to raise animals. If I had the money I would buy land so I could do this.

The differences in the long term goals of these two Real Farmers have no impact on their current commitment to Ntimane and to the social institutions existing in Ngwempisi, but this will change according to circumstances. I concluded that the base level of common interest for the Real Farmers is that of their conjugal families.[1] Because Swazi social organization is based on polygyny, the Real Farmers additionally share a common interest in a mutual commitment to monogamy and conjugality. Their solidarity in Ngwempisi is based around satisfying the need for a system which favours their conjugal families.

If the primary focus of Real Farmer interests is their conjugal families, with a resultant diminution of the importance of lineage and locality, other principles govern their relationships with the members of the general community. For example, I inferred that the mechanisms by which the Real Farmers exercise power, sustain the redistribution of cheap food (vegetables), and stress adherence to community norms and values, could constitute them as a 'moral' economy (Scott, 1976, p.3). At the centre of Scott's moral economy thesis is the notion that the individual was forced into patterns of reciprocity to guard against the inevitability of what Sahlins (1972) refers to as 'house hold failure', wherein '... a fair percentage of domestic groups persistently fail to produce their own livelihood, although organized to do so' (p.69).

In Ngwempisi, the Real Farmers retain their economic advantage by sustaining local sales through the over supply of cheap vegetables to the local market. They also further distribute largesse through social activities such as the work parties, by acting as hosts and patrons, and allowing access to scarce items such as vehicles. Strong traditional imperatives support this process. For example, the institution of begging facilitates the growth of patron client relationships, and reduces evidence of sharp class cleavages in the area, by maintaining a reasonably equitable distribution of basic commodities. The strategic under use of land at the homestead by the Real Farmers is also in line with a similarly cautious approach to land use among the general population. Finally, the retention of strong kinship ties by the Real Farmers, despite their focus on the conjugal family, ensures the continuity of networks, and safety nets by which redistribution and reciprocity function and patronage evolve.

Although strong elements of a moral economy are evident in Ngwempisi, the activities of the Real Farmers, in limiting general access to the resources of the chiefdom and seeking ultimately to depopulate the area, demonstrate a more fundamental motivation for their activities. By espousing moral economic values, the Real Farmers gain individually, and thus act to influence social norms in a manner which parallels that which Popkin (1979) describes as being '... malleable, re-negotiated, and shifting in accord with considerations of power and strategic interaction among individuals' (p.22), in a situation where entry to the market does not involve the diminution of rural social institutions (Popkin, 1979, p.267). The moral community does impose limits on the freedom of individuals to pursue their own course. For example, in Ngwempisi the Real Farmers continue to be concerned with avoiding witchcraft, and so are continually forced to reassess their actions in that light.

The activities of the Real Farmers can be conceptualized as the reconciliation of short term and long term strategies. In the short term, by reinforcing and increasingly dominating local institutions, the Real Farmers safeguard themselves, and their interests, against the consequences of alienating their kin and neighbours. In the long term, they gain control over, and restrict access to, the resources of the chiefdom. In this sense, individuals make choices about their values and actions and so change their own society (Harriss, 1986b, p.21).

In restricting access to resources, the domestic group, which is subject to constant redefinition, emerges as the apogee of Real Farmer self interest. Although monogamous conjugality forms a model domestic grouping, I have recorded viable variations. For example, the upsurge in female Real Farmers and in the number of unmarried mothers with non-filiated children, all of whom form basic domestic units, indicate that the domestic group cannot be rigidly defined. Particularly noteworthy is the fact that domestic groups, such as those formed by a single woman and her children, involve the reduction of self interest to the level of the individual, such that each person perceives his or her interests to be concurrent with those of the wider community.

Therefore, I conclude that the strategies of the Real Farmers as a group, in upholding custom in the short term to achieve particular goals in the long term, are coordinated and cohesive because the common short term goal is readily identifiable. The Real Farmers want proprietorial control over their homestead land to serve the interests of their conjugal family. As individuals, the Real Farmers confront the state, which controls land and is represented by the Chief and others, and, by using the symbols of custom, negotiate a new division of resources.

The community and the state

Visible in the reconciliation of short term and long term strategies is the intimate nature of the relationship between the community and the Swazi state. Over time this relationship has manifested itself in a variety of forms, dependent on political circumstances, but has remained consistent in other ways. For example, there has always been a lack of material differentiation between those at the highest levels of the state and the commoners. A differential possession of 'high' culture (Fallers, 1961 p.110) has never existed in a manner that suggests the cultural and economic domination of the hinterland, or some manner of extraction of surplus to the urban areas. Rather, 'high' culture was traditionally, and is still, predominantly, expressed as a rural ideal, attested to by the maintenance of rural homesteads by prominent members of the ruling aristocracy and the urban elite. While competition for rural land exacerbates idealization of rural dwelling among aspirants, at a basic economic level rural agriculturalists enjoy comprehensive tax benefits over 'pay as you earn' (PAYE) urban wage and salary earners.[2] Therefore, the Real Farmers enjoy both a cultural and an economic advantage in commercial farming on Swazi Nation Land, because the taxation system favours them, and they also live the contemporary Swazi ideal as economically self sufficient rural dwellers.

The intimacy of the relationship between the community and the state is further evidenced by the effect of the Rural Development Areas Programme in Ngwempisi. Despite the differences between what was anticipated and what has occurred, the rapid emergence of the Real Farmers, as a result of the programme, demonstrates the degree to which government actions affect the rural community. The view of the rural community in its relationship to the state is, in some aspects, at odds with those views of the community I described in Chapter One. For example, the dualist view of Swazi society used in the formulation of the Rural Development Areas Programme, in which tradition is seen as an obstacle to development, is flawed when the complexities of rural urban relationships through the rural homesteads are considered. To the contrary, I would argue that ideologically, tradition, as a source of legitimacy, has proved vital in the process of change. Far from inhibiting change, tradition is demonstrably the crucial element in comprehensive and non-traumatic change. This is a view reflected in an array of historical and anthropological studies carried out in Africa in the 1940s and 1950s (Barnes, 1954; Fallers, 1965; Nadel, 1964; Smith, 1960), in which change involves the acceptance of cultural elements by adding, rather than substituting, these elements for those already in existence (Herskovits and Bascom, 1962, p.6).

The complexity of social, economic and political relationships between the rural homesteads and the urban areas further denies any sense of a unilinear progression from tradition to modernity. As I stated in Chapter One, this notion was a fundamental assumption in the formulation of the Rural Development Areas Programme. The Real Farmers, through their ability to support tradition while achieving their own goals, demonstrate conclusively that tradition and modernity are concepts which they use in discourse, but in action are not separated.

The dualist view of Swazi society used in the formulation of the Rural Development Areas Programme incorporated some complementary theoretical factors of interest. For example, public policy analysis (Higgott, 1983, p.26), focusing specifically on order and regime maintenance, can provide useful insights into Swazi political dynamics. Order and regime maintenance are not sustained through direct bureaucratic action, but rather through the growth of indigenous layers of people which support the state. As one such layer, the Real People, and especially the Real Farmers, become ever more reliant on the infrastructure of an increasingly powerful centralized state. In Ngwempisi, this is clearly evidenced by the circumstances I described, in which the farmers have come to depend on hybrid seeds, chemical fertilisers, roads, markets, irrigation technology and many other services. Despite my conclusion that the Chief is still the most important political figure to the Real Farmers, this is only in the short term. Ultimately, the Real Farmers know that the issue of land scarcity will be solved by legislative action by the central government, and that they must satisfy the demands of this government. In the long term the Real Farmers must support the state to continue their success, which is indeed what is happening.

As I discussed in Chapter One, views of contemporary Swazi rural society have come from a range of perspectives, and have incorporated strong criticisms of development policy and practice. The strongest criticisms can be attributed to dependency theorists. In the broad sense of dependency (Higgott, 1983, p.48), the Swazi data indicate that the inherited colonial structure has not blocked capitalist development, though elements of such may be perceived in aspects of land acquisition and labour use in agri-business operations in the low veldt. The Real Farmers are an example of a form of capitalist development, one which has emerged because of the conditions created by colonialism, and through inherited colonial structures.

The suggestion that colonial structures have blocked capitalist development is a view which overlooks the active role of the King and the ruling aristocracy in restricting capital penetration into Swazi Nation Land.

The King used his company, Tibiyo Taka Ngwane, to launch joint ventures with foreign capital on selected areas. This process restricts the movement of capital into the Swazi heartland, except for those transfers which occur through the remittances and investments of domestic groups. At the same time, outflows of commodities from Swazi Nation Land were thus restricted to agri-business concerns, with profits accruing to Tibiyo Taka Ngwane.

The effect of this restriction of capital penetration in Ngwempisi has been to provide an ideal environment for capitalist development. Taxation levels are low, making commercial farming economically attractive. The distance to markets makes local sales by the Real Farmers attractive to customers. The situation is such that the Real Farmers have no competition from large scale producers. Because of the attractiveness of local sales, the Real Farmers are able to enter the external markets very cautiously, without constantly having to gamble on making sales. The lack of comprehensive capitalist development in Ngwempisi also provides some subsistence security for those who labour in the towns. Such people are able to enter and leave the labour market without risking their livelihood. As I have shown, the rural homestead is a safety net and a resource. A disgruntled worker can either return to the country, or stay with an urban patron.

The dynamic nature of local economic growth does lend some credibility to a 'productionist' perspective (Higgott, 1983, pp.62-65), since an 'articulation' of capitalist and non-capitalist (or pre-capitalist) modes of production gives shape to particular social formations. In particular, the Swazi data seem to indicate some validity in Rey's notion of a 'process' of three stages of articulation, whereby initial linkages in exchange lead, through interactions with capitalism which reinforce pre-capitalist modes, to an eventual total subordination of pre-capitalist modes (Foster-Carter, 1978, p.56). In the context of Swaziland, initial contact with capitalism reinforced pre-capitalist modes of production as the need for labour on the South African mines sustained the polygynous female dominated homestead economy, in the same way Swazi militarism had done previously (see Chapter One). The second stage of articulation, now occurring, involves phenomena such as Real Farmer activity utilizing and eventually subordinating the pre-capitalist mode. In Ngwempisi, the Real Farmers use traditional agricultural techniques at their homesteads, and engage in significant traditional political activity within the chiefdom, with a long term goal of supplanting that system. The third stage, the total disappearance of the pre-capitalist system, has not been reached in Swaziland except on Individual Tenure Land.

However, these practices have only been reinforced in particular contexts, based on truncated and notional perceptions of pre-capitalist modes, which historical evidence suggests are the result of substantial syncretic processes. As Higgott (1983, p.65) notes, the value of a productionist analysis is its ability to highlight ' ... the complexity and array of relationships which are possible under capitalism'.

I have argued that the majority of the commentators on the nature of contemporary Swazi society have been reluctant to support dualistic views of Swazi society (see Chapter One). I suggested that most of these researchers focused on local data indicative of class structures related to, or emerging because of, social change. Among these researchers are Russell and Ngubane, both of whom have argued strongly that the rural homestead is highly adaptive to the demands of the capitalist economy. Both also argue that the rural homestead retains its economic integrity in the face of capitalist activity. Russell suggests that capitalism has become part of the homestead (1983, p.319), while Ngubane proposes that this same process occurs through cultural integration, an occurrence she believes to be common in Swaziland and in Africa generally (1983, p.3).

In Ngwempisi, my data strongly support those of Russell and Ngubane. The Real Farmers, among others, use the rural homestead selectively to enter and leave the market. Also, the rural homestead extends into urban areas as the wide ranging networks of patronage and remittances operate within kin groups. This is illustrated by the examples I gave of planned migration to the city by wage and salary earners, who use the resources they find there to maintain, or establish their rural homestead. In some cases, wage and salary earning satisfy a short term need for capital, which is then used to develop a commercial enterprise at the homestead.

The strength of Real Farmer activity in the face of government officials indicates a special relationship between the Ngwempisi community and the Swazi state. I have described a situation where local government officials do not constitute a threat to the power of the Chief and his Real People. I have argued further that tradition has proven crucial in the form social change has taken, and that this process has occurred in spite of the Rural Development Areas Programme, but also because of it. The relative powerlessness of government officials to enforce the programme has been significant throughout the process. As a result, the circumstance has not emerged where the lack of farmer enthusiasm for the Rural Development Areas Programme can be seen as a form of peasant resistance, or as agriculturalists taking the exit option. As such, the rural agriculturalists cannot be seen as an 'uncaptured' peasantry (Hyden, 1980). Hyden's context was that of the Ujamaa socialist experiment in rural development in Tanzania, a programme in marked contrast to the Rural Development Areas Programme in Swaziland, although based on similar views of contemporary Tanzanian society.

Ujamaa socialism was based on the writings of J. K. Nyerere (1968, 1969, 1971, 1973, 1975, 1978, 1979a, 1979b, 1979c), the then President of Tanzania.[3] Nyerere implemented Ujamaa stressing the non-coercive nature of the policy (Duggan and Civille, 1976). In the face of public indifference, he subsequently instigated Ujamaa *vijijini* (villagization), eventually resorting to draconian measures in an attempt to achieve the Ujamaa goals (Boeson, Madson and Moody, 1977, pp. 150-151; Bowen, 1983; Coulson, 1982, p.262; Mwapachu 1979, p.116; von Freyhold, 1975, p.75). Nyerere retired voluntarily in 1985, publicly critical of his own policies, despite improvements in literacy and basic medical facilities (Manley,, 1986; Onim,, 1985a, 1985b; South, 1985). He ceded power to Zanzibari businessmen politicians who have set Tanzania on a new economic course based on private enterprise strategies (Onim, 1985b, pp.21-22; Sheunyane,, 1985, pp.23-26).

The Ujamaa period of the 1970s in Tanzania was marred by bureaucratic and governmental corruption, wasted resources and poorly planned programmes. After an initial period, attempts by government officials to enforce planting routines, resettlement programmes and centralized marketing programmes, led to growing passive resistance, as bureaucratic inefficiency and corruption became evident to the producers (Coulson,, 1979, 1981, 1982; Hyden, 1980; Ingle, 1975; Sheunyane, 1985). In areas closest to regional centres and under the constant scrutiny of officials, the capacity of agriculturalists to feed themselves was actually reduced, and higher incidences of malnutrition were noted among those who participated in the various programmes (ten Raa, 1983a, 1983b).

In Swaziland, under the Rural Development Areas Programme, government officials have none of the coercive power of the Tanzanian officials. This relative lack of power has made it possible for farmers to be very selective about the services they choose to use, and has restricted the ability of officials to pursue particular goals. In their dealings with the Real Farmers, officials are limited to advising farmers without being able to enforce their recommendations. Some officials confided to me that they found this situation frustrating.

The essential difference between Ujamaa and the Rural Development Areas Programme, aside from their scope and differences in detail, is the coercive power of government officials. The Tanzanian officials received their power directly from the president. They were directly responsible to him and they carried out his wishes (Maro and Mlay, 1979, p.296). This understanding of the president's direct role was well understood by the general public (Bakula, 1971; Mashauri, 1971; Musoke, 1971; Ntirukigwa, 1971). In Swaziland, the King's direct influence over the Rural Development Areas Programme is far less obvious. I encountered one specific reference to the King's wishes, and this was at the Mzimneni Resettlement Scheme. In Chapter Two I described this comment as indicating a ploy by some officials to encourage people into resettlement by using the King's name. Farmers told me that similar things had happened elsewhere but they were not sure of the details. I concluded that the King did not allow his officials any coercive power, and he did not himself intervene directly in the routine of the Rural Development Areas Programme. Nevertheless, as I will show, rural development was essential to the King in the post independence era.

Swazi corporatism

Higgott (1983) suggests a model which may enhance the understanding of the relationship of the community to the state in Swaziland. In particular, this model may help in providing an understanding of the relative success of the Real Farmers throughout the Rural Development Areas Programme period. The coercive power of government officials is the crucial issue, a power which is not held by Swazi rural development officials because they had not been ceded it by the King.

Formerly, the state in Swaziland was perceived by non-Swazi very much in the '... pluralist sense of "neutral arbiter"' (Higgott, 1983, p.78), or as having '... only a shadowy existence at best merely an instrument which is used by this or that class or alliance of classes' (Randall and Theobald, 1985, p.135). This view of the Swazi state envisaged it only as the government of Swaziland, and failed to account for the evidence of the Swazi state as part of a larger Swazi nation. This latter concept was evidenced by governmental concerns for reunification with the large number of ethnic Swazi who live in the Republic of South Africa (Government of Swaziland, 1982a, 1982b, 1983a).

The process of development policy formation and implementation, and responses of farmers to that policy in Swaziland, reflect a complex relationship between the community and the state. The contemporary Swazi state takes power from public acceptance of the legitimacy of government.[4] This factor is particularly important, especially to the ruling aristocracy. Also important is public acceptance of the right of the state to use force, and to uphold its claim to binding rule making over a defined territory, by virtue of commanding a monopoly over the legitimate use of force (Gould and Kolb, 1964, p.690; Mann, 1983, p.373).

I have argued that state legitimacy is not in question in Ngwempisi, because people perceive the government of the prime minister and his cabinet to be different from, and inferior to, that of the King. The King's absolute right to rule has never been an issue in the Ngwempisi community, and his right to use force to maintain his government is not questioned. Ironically, the power of the ruling aristocracy, in anticipation of any questioning of that legitimacy, depends increasingly on its direct economic control over the Swazi Defense Forces.[5] So, despite his unquestioned right to grant coercive power to his rural development officials, King Sobhuza II did not do so. In fact, as I have suggested previously, he obstructed the acquisition of power by these officials by giving the chiefs powers of veto and direct access to regional officials. The only possible conclusion to be drawn is that rural officials did not have coercive power because the King did not want them to have it.

In contradiction to his stance on power for rural development officials, in the industrialized sector of the Swazi economy the King allowed his officials real coercive power, particularly in the area of industrial disputation. Unrest among labour has always been dealt with relatively harshly. The King and his government were intolerant of unionism throughout the 1970s. For example, a march of railroad workers in 1975 was met with tear gas and mass arrests, as was a strike by sugar workers at Big Bend in 1978. Teacher and student protests over wage levels in 1975 and 1976 escalated to the point where the police where authorized to fire on demonstrators (Booth, 1983, pp.75-76). These examples are only a few of the occasions in which unrest was strongly reduced by the King's use of force. If he wished, Sobhuza had the ability to dispense coercive powers to all his officials, but he chose to do so quite selectively.

The Swazi state exemplifies that model proposed by Higgott and Robison (1986b), who argue that in some of the more rapidly developing countries of the Third World, it is the '... corporatist form of government' which is increasingly evident, and that the attraction of this form is its ability to accommodate laissez faire growth theory on the one hand, and demands for political order, elite security, and regime maintenance on the other (p.43).

> In essence, corporatism is a system of politico-economic organization in which the state directs and controls predominantly privately owned businesses according to four principles: unit, order, nationalism and success. In such a context, corporatism exhibits a tendency to over-control at the political level and under-control at the economic level. The aim is to provide a fairly unrestricted degree of freedom to a society's entrepreneurial groups while providing strong control over the workforce - especially with regard to wage rates and union activity (Higgott and Robison, 1986b, p.43).

Randall and Theobald (1984) argue that the broad model of corporatism entails a continuum, one extreme being spontaneous organization, in which the state's role is arbitration, and the other extreme being state corporatism, in which power and initiative are concentrated in the state itself (1984, p.173). This can be further divided into two forms; 'inclusionary', in which the state uses controlled mobilization and co-option to forestall the independent development of worker and peasant organizations; and 'exclusionary', in which already formed organizations are deliberately demobilized (Randall and Theobald, 1984, p.173). Within the planned economy, further distinctions are those of the 'indicative' style, characteristic of mixed economies, and the 'imperative' style usually associated with a totalitarian regime. (Robertson, 1984, pp.62-64).[6]

In many respects the Swazi state exhibits corporatist characteristics. Inclusionary and exclusionary tactics have been used. Funding and access to resources have always been oriented towards government initiated rural development schemes in a manner suggestive of an inclusionary tactic. Ntimane is no exception. To receive funds, Ntimane has to be part of a rural development area, and has to put these funds into a government approved project. Exclusionary tactics were also evident. For example, unions and political parties are banned under the threat of sixty day detention orders. This activity only occurs in either the towns or the agri-business complexes; such tactics are never applied to agricultural cooperatives in rural areas.

Differences also occur in the style adopted by government officials. In country areas, officials, because of their lack of power, use indicative tactics in their efforts to motivate and lead farmers. As I described previously, any attempt to use imperative tactics incurs only anger and opposition from the farmers. This was well demonstrated by the example of the drunken official whom none of the prominent Real Farmers would meet (see Chapter Six). However, in the towns and on the agri-business estates, government officials and management (because of government support for business) are able to adopt imperative techniques with impunity. It is a common complaint among workers that, through the *ndabazabantu* system (with the workforce being represented by an appointee of the King), they are not allowed to participate in negotiations, but are simply informed as to what their working conditions will be.

The imperative style is also adopted by government leaders, cabinet ministers and the prime minister, to pronounce on any particular social issue. These pronouncements are accepted, but many people expressed dismay to me over what they perceived to be arrogance on the part of such people. There is an acute awareness among urban workers that the imperative style can be backed by police action.

In Ngwempisi, the use of the indicative style of inducing change requires further qualification, lest the relative success of this approach be envisaged as a validation of a demonstration effect. The Real Farmers have not responded enthusiastically to government initiatives, but have made individual choices about perceived opportunities. This lack of enthusiasm is even more striking when it is considered that most Real Farmers came to farming from wage work in intensive commercial industries in rural and urban areas, where they have often worked in technically sophisticated industries. Despite this experience, they have reacted cautiously to new technology. In my opinion this caution relates more directly to considerations of the cost effectiveness of the technology than to the obstinacy of the farmers.

Earlier, I mentioned the farmer Vilane, who is a tenant on a large private farm which borders Ntimane, a few metres across the Mtimane. In exchange for some labour, the private farmer allows Vilane, and his other tenants, the use of land for farming and grazing. Vilane engages in commercial farming, and grazes fifteen cattle. In conversation with me he enthusiastically endorsed the farmer's emphasis on under grazing, and told me that such an approach to stocking was crucial for preserving grazing on Swazi Nation Land.

Vilane was convincing in his advocacy of limited grazing herds. Therefore, I was surprised to hear him state that, once he had reached his allowed limit of fifteen head on the private farm, he entered into cattle lending arrangements with his kin on Swazi Nation Land, thus aggravating the very condition he criticizes. Furthermore, when I asked Vilane to tell me about his ideal method of farming, he cited Ntimane, not the employer he so admires. The ideal factors he cited were social factors such as cooperation and a strong organization. The goals he admires are education and rural living.

Therefore, the success of the indicative method of inducing change is limited to farmers' perceptions of the worth of the technology, rather than to their natural reluctance to use new technology. For this reason, using imperative tactics would compound the inevitability of their resistance to technical innovation. The farmers judge the worth of the technology in many ways which are social, rather than economic. I have shown that technologies to increase production at Ntimane are worthless while current levels of surplus occur and external market opportunities are limited.

The lower rate of technology adoption at the homesteads would indicate that people perceive few benefits in using such technology. Yet I described how quickly expensive and economically non-productive stoves have been incorporated into homestead kitchens. This indicates a specific and localized awareness of what constitutes economic rationality, as the Real Farmers' need for significant arenas of political action rapidly justifies the outlay.

State control and the Real Farmers

Earlier in this chapter I suggested that one of the reasons why the Real Farmers support the state, as represented by the King and his chiefs, is because they know that ultimately the issue of land tenure will be decided at that level. Furthermore, as I have argued throughout the thesis, the Real Farmers are moving into a relationship of greater dependency on infrastructures which are centrally controlled.

This acceptance of their dependence by the Real Farmers contradicts the view, described by Higgott (1983), that governments become authoritarian because they cannot achieve such dependence in their farmers. Such governments also strive to maintain centralized administrative control, rather than allow diffusion of administrative structures and power at a local level. Again this is in contradiction to the situation I have described, where the Chief and the Real People hold significant power at the local level (Higgott, 1983, pp.85-86,98).

King Sobhuza II held almost complete power in Swaziland. He promoted rural development through the Rural Development Areas Programme, but when that programme began experiencing difficulty achieving set goals, he did not resort to authoritarianism. Throughout the process of rural development in Ngwempisi, and despite the Rural Development Areas Programme, the King allowed the Real Farmers to emerge and to share the Chief's power.

Why did the King truncate the power of Rural Development Areas Programme officials in Ngwempisi by leaving all real power with the Chief? When the Rural Development Areas Programme began to have difficulty in achieving set goals, why did the King choose not to resort to authoritarianism? Finally, why do the Real Farmers pursue a future of increased dependence on the infrastructures of the state?

The answers to these three questions lie in the King's motives in welcoming formal rural development programming to Swaziland. By the short term use of the Rural Development Areas Programme, the King achieved the long term goal of centralized power, restricted to those who give him their unqualified support. Sobhuza II left power with the chiefs because he never intended to let real power move away from the ruling aristocracy, of which the chiefs are members. The lack of authoritarianism occurred because the King needed foreign capital in the post independence years to provide employment and opportunity for the growing urban population which was becoming restless, not because he wished for a successful rural development programme. I contend, therefore, that the King knew that aid monies were only going to be available from foreign governments if they were tied to rural development. This had certainly been the case with aid money from the United Kingdom. Others have described this as a situation in which the rhetoric of development and the implementation of the programme in effect form the basis of communication between the indigenous state and the international community (Higgott, 1983, p.xiii; Morss and Gow, 1985, p.xiv; Morss and Honadle, 1985, p.202), and as such involve expressions of generally recognized principles of universal need.[7]

The final question of why the Real Farmers are prepared to move into a relationship of greater dependence on the infrastructure of the state is at the crux of the entire issue of community state relations in Swaziland. The Real Farmers trust in their ability to comprehend the political process. Because they share power at the local level, and because they can see the linkages with power at the national level, they believe they can anticipate the vagaries of government, as it is exercised through the control of the ruling aristocracy. The Real Farmers recognize their integral role in the sustenance of royal power.

For the moment, royal power rests on public acceptance of the legitimacy of the institution of kingship. King Sobhuza II understood this reality, and he knew that he had to maintain that legitimacy despite the radical changes which were occurring over time, throughout his domain. Sobhuza realized that he needed to become the final arbiter of tradition in the minds of his people so that he might reform tradition, as needed, to retain his legitimacy as King. He started this process, in 1934, by inviting the anthropologist Hilda Kuper to work in Swaziland. At the time he stated:

> Anthropology makes possible comparison and selection of lines of further development. European culture is not all good; ours is often better. We must be able to choose how to live and for that we must see how others live. I do not want my people to be imitation Europeans, but to be respected for their own laws and customs (Kuper, 1978, p.2).

Sobhuza understood perfectly the role of interpretations of tradition which were deemed legitimate by his people. Anthropological research, sympathetic to the aspirations of the King, provided an orthodoxy which supported Sobhuza in his role as ultimate arbiter of tradition, and formed the conceptual basis for all the development oriented research carried out from the 1960s onward. The seminal role of Hilda Kuper, in sustaining Swazi traditional orthodoxy over the years 1935 to 1983, must be recognized, as must the degree to which her ethnographies form the conceptual framework for most subsequent analyses of Swazi social organization.

Sobhuza propagated a mythology of harmony and equilibrium by re-instituting public rituals such as the *umhlanga* (reed dance) and the annual *incwala* (first fruits ceremony). He sustained a model of traditional Swazi society in the minds of his people, which Kuper placed in an appropriate academic context.

Throughout his reign, the King sought to consolidate his own position in the colonial administration. His main technique during the period was to adopt a deliberate policy of technological and cultural integration, though he always maintained that this policy was to preserve Swazi tradition, not to destroy it.

The King's endorsement of deliberate and legitimate cultural integration subsequently permeated every level of Swazi society, and facilitated the ready incorporation of foreign technological and cultural elements without any sense that tradition was being lost.

Sobhuza was an exceptionally gifted and astute politician and, by his actions in propagating selective change, he imbued the process with a high moral value. He added credibility to this moral process by constantly referring back to the past, and by idealizing the process by which the Swazi had maintained their state through negotiation rather than confrontation. Rather than resisting established colonialism, Sobhuza II sought to advantageously position himself within the colonial hierarchy. By working within the colonial administration, Sobhuza gave the authorities no opportunity to alienate him from his people, Instead he became the intermediary for his people and kept himself constantly on the political stage.

When the King finally gained absolute control of the state after 1973, his experience was such that he was able to manage a government structure of great complexity, without his people questioning the appropriateness of his role.

His political strength, after the 1960s, lay in his ability to dichotomize the settlers on Individual Tenure Land and the bulk of the population on Swazi Nation Land, in a way that achieved harmony, since those in each sector had a clear understanding of their rights. Only the growing urban population required firm control, and this he initially achieved by silencing industrial unrest and political opposition. With the advent of Independence, Sobhuza had the requisite political and literary resources to reinforce aspects of his traditional model in a manner attractive to the modernization theorists representing foreign governments. Subsequently, he used the funds which became available through the various development programmes to employ the disenfranchized educated Swazi, to reinforce his position among his Swazi Nation Land power base, to maintain commercial opportunities on Individual Tenure Land, and to develop his own financial power base in Tibiyo Taka Ngwane through joint ventures with foreign capital.

In achieving this result, Sobhuza brought to fruition a long term process of integrating popular conceptualizations of tradition and modernity in a manner which reinforced his own position in the colonial bureaucracy, and left him well placed to completely dominate the independence process, as indeed he did.

The King used the Rural Development Areas Programme to facilitate the growth of indigenous and supportive social formations such as the Real Farmers. By treating aid as a compensation, and therefore part of the national estate, he was able to employ the urban population who were so restless during the 1970s, and so gave himself the breathing space needed to allow two developments. First, the ruling aristocracy now have their economic base established through the Tibiyo Taka Ngwane corporation, and their military base through the Swazi Defence Force. Second, because he did not risk disrupting existing patterns of production on Swazi Nation Land, the King solidified his traditional political power base in rural areas. The Real Farmers, therefore, represent the best interests of the King. They are rural based and politically conservative. They distrust that sector of the government represented by the urban elite, but have faith in that sector represented by the ruling aristocracy.

In conclusion, in this chapter I have reiterated my contention that the Real Farmers uphold tradition in a system which they ultimately seek to change. I have further argued that there is a direct relationship between the activities of the Real Farmers in Ngwempisi, and the nature of the relationship between the local community and the state. In the same way that the Real Farmers uphold tradition to achieve their long term goals, so too King Sobhuza II upheld tradition throughout his reign so that he could eventually dominate the post independence Swazi state. Furthermore, the King used the funds available through the Rural Development Areas Programme to provide employment opportunities for the growing urban population, which was a threat to his position in the post independence years. By this action the King gave himself the breathing space necessary to develop his power base on Swazi Nation Land. He was successful in this endeavour, for over the period of the 1970s and the 1980s, people with a vested interest in the contemporary political system, such as the Real Farmers of Ngwempisi, have emerged to fill that crucial role.

Conclusion

In Ngwempisi, the Real Farmers dominate processes of change by upholding tradition. In this book I have described this process, and have shown how the Real Farmers use the language and symbols of custom to facilitate change. They are united in this by their common goals, the most immediate of which is their determination to gain proprietorial control over land they currently use by right of usufruct.

The activities of the Real Farmers are described in two major parts. The first part dealt predominantly with the effect they are having on the economy of Ngwempisi. I argued that the Real Farmers are dominating the local economy by manipulating access to, and the use of, land, labour, and finance, in a manner that ultimately gives them increasing control over that economy.

The second part dealt with the impact of Real Farmer activity on the social organization of Ngwempisi. I described a process by which the Real Farmers uphold certain customary values related to lineage and locality, and power and authority, but actually favour social relationships that encourage monogamy and conjugality, and the rights of individuals over the rights of lineage members. Importantly, the Real Farmers, by supporting the retention of power and authority by the Chief of Ngwempisi, have come to share that power and authority in a way that reinforces social change in line with their values.

In the later stages of the book data were presented suggesting that the Real Farmers do not act in isolation in Ngwempisi, but have emerged in a wider process of social differentiation involving the whole nation. This differentiation has local implications in terms of restricting access to the general resources of the community, and national implications in terms of social organization in an urban environment. In Ngwempisi, the Real Farmers differentiate people on the basis of their commitment to particular values, either to enhance or reduce their access to land, labour and finance. At a national level, the Real Farmers use urban opportunities to extend their rural homesteads into the cities so as to establish their commercial operations on Swazi Nation Land.

In the final stages of the book, the argument was presented that the Real Farmers are part of a complex transformation which has seen the re-emergence of royal control over the Swazi state. The crucial factor for the royal family, as for the Real Farmers, has been the scarcity of land brought about by the colonial political system. I suggested that the techniques by which the late King Sobhuza II gained control, and consolidated his power, in Swaziland, were very similar to those used by the Real Farmers in Ngwempisi. Like the Real Farmers the King upheld custom in such a way as to become indispensable in the colonial period, both to the colonial authorities and his own people, for whom he was the final arbiter of tradition. With Independence, he had the moral authority to vie for and ultimately seize, power. Like the Real Farmers his actions can only be judged in the long term, and indeed after his death, royal power, in the hands of his son, is stronger and more comprehensive than ever before.

Like the late King Sobhuza II, the Real Farmers plan for an indefinite and uncertain future. They plan with finesse and vigour, they challenge the existing order and use their influence with confidence and rectitude. Nevertheless, they recognize their own limitations, and their relative vulnerability, and so they draw about them the reassurances of custom, which they express to themselves, and to others, through images of the past.

Notes

1. Ideally, the conjugal family constitutes the basic unit of production and self interest. But as the examples of female managers indicate, in the

absence of Real Farmer males, female 'houses', conforming to a more traditional model, constitute a basic unit of consumption.

2. Individual taxes comprise 11.7 per cent of total revenue (Government of Swaziland, 1985, p.62), and of that very little comes from small farmers on Swazi Nation Land (Dlamini, 1984; Nkambule, 1984; Simelane, 1984).
3. Nyerere was specifically influenced by the work of the Catholic theologian Teilhard de Chardin as well as by Maoist thought as it related to a rural based revolution, the status of women, and the role of the family (Bowen, 1983; Duggan and Civille, 1976, p.46; Han, 1972, pp.108-109; McLellan, 1975, pp.45-46).
4. Theodorson and Theodorson define the state as: 'A political form of human association by which a society is organized under the agency of a government which claims legitimate sovereignty over a territorial area, authority over all of the members of the society, and the right to use physical force when necessary to ensure the effective exercise of its legitimate control' (1979, p.412).
5. The Swazi Defense Forces are paid directly by the Tibiyo Taka Ngwane corporation.
6. The indicative style is one in which emphasis is placed on transmitting ideas by suggestion, as opposed to the imperative style, whereby ideas are transmitted by command.
7. These principles are known as the Basic Needs Approach and are said to define certain universal needs basic to all people. These needs are to be met through a redistribution of material wealth and government resources, and take the form of hospitals, clean water, schools, roads and other infrastructure (Conyers, 1982, p.28; Higgins and Higgins, 1982, p.32; Webster, 1984, p.35).

Bibliography

Allan, W. (1949), 'Studies in African Land Usage in Northern Rhodesia', *Rhodes-Livingston Papers,* vol. 15.

Allan, W. (1965), *The African Husbandman*, Oliver & Boyd, Edinburgh.

Armstrong, A. (1985), 'Legal Aspects of Land Tenure in Swaziland', USAID/University of Swaziland, Kwaluseni.

Armstrong, A. & Nhlapo, R. T. (1985), *Law and the Other Sex: The Legal Position of Women in Swaziland*, USAID/University of Swaziland, Kwaluseni.

Armstrong, A. & Russell, M. (1985), 'A Situational Analysis of Women in Swaziland', UNICEF\Social Sciences Research Unit, Kwaluseni.

Bakula, B. B. (1971), 'The Effect of Traditionalism on Rural Development: The Omwinazi Ujamaa Village, Bukoba' in Proctor, J.H. (ed.), *Building Ujamaa Villages in Tanzania*, Tanzania Publishing House, Dar es Salaam.

Banford, W. R. G. (1964), 'Demography: Education' in Holleman, J. F. (ed.), *Experiment in Swaziland*, Oxford University Press, London.

Barker, D. (1965), *Swaziland*, Her Majesty's Stationery Office, London.

Barnes, J. A. (1954), *Politics in a Changing Society: A Political History of the Fort Jameson Ngoni*, Oxford University Press, London.

Basham, R. (1978), *The Cross-Cultural Study of Complex Societies*, Mayfield Publishing Company, Palo Alto.

Beemer, H. (1935), 'The Swazi Rain Ceremony: (Critical comments on P. J. Schoeman's article)', *Bantu Studies*, vol. 9, pp. 273-280.

Beemer, H. (1937), 'The development of the military organisation in Swaziland', *Africa*, vol. 10, pp. 176-204.

Billard, A. (1986), 'Some South African refugees in Swaziland', *Refugees*, vol. 2, pp. 12-13.

Black-Michaud, A. (1981), 'Homestead vs Households When Collecting Socio-economic Data', Ministry of Agriculture and Cooperatives, Mbabane.

Boeson, J., Madsen, B. & Moody, T. (1977), *Ujamaa: Socialism from Above*, Scandinavian Institute of African Studies, Uppsala.

Bohannan, P. (1959), 'The impact of money on an African subsistence economy', *The Journal of Economic History*, vol. 19, pp. 491-503.

Bonner, P. (1983), *Kings, Commoners and Concessionaires: The Evolution and Dissolution of the Nineteenth-Century Swazi State*, Raven, Johannesburg.

Booth, A. (1983), *Swaziland: Tradition and Change in a Southern African Kingdom*, Westview Press, Boulder.

Booth, A. (1986), 'Homestead, State and Migrant Labour in Colonial Swaziland' in Daniel, J. & Stephen, M. (eds.), *Historical Perspectives on the Political Economy of Swaziland*, Social Sciences Research Unit, Kwaluseni.

Bowen, P. N. (1983), *A Time for Planting and a Time for Harvesting: J. K. Nyerere and the Dynamics of Ujamaa Socialism in Tanzania*, unpublished honours dissertation, The University of Western Australia.

Cappetta, M. (1983), 'Population, Food and Nutrition: Swaziland 1940-1982' in De Vletter, F. (ed.), *The Swazi Rural Homestead,* Social Sciences Research Unit, Kwaluseni.

Chayanov, A. V. (1966), *The Theory of Peasant Economy,* Richard. D. Irwin, Homewood.

Cartwright, J. (1978), 'The Lion of Swaziland' in Isaacs, A. & Jones, P. (eds.), *The World Around Us*, BBC Colour, London.

Cohen, D. L & Daniel, J. (1981), *Political Economy of Africa*, Longman, London.

Cohen, N. P. (1983), 'Recurrent Budget Analysis for Irrigation Projects'. United States Agency for International Development, Mbabane.

Cohen, N. P. (1984a), 'Estimates of per captita/per houschold income in rural and urban areas', *Econews*. vol. 2, no. 2, pp. 1-18.

Cohen, N. P. (1984b), 'Swazi Debt Situation and the Cyclone', United States Agency for International Development, Mbabane.

Cohen, N. P. (1984c), 'Forecast of Economic Trends', United States Agency for International Development, Mbabane.

Cohen, N. P. (1984d), 'Tibiyo Taka Ngwane', United States Agency for International Development, Mbabane.
Cohen, N. P. (1985a), 'The World Bank's Report Economic Mission to Swaziland: Back to Office Report/Issues Paper', *Econews*, vol. 11, no 9, pp. 1-10.
Cohen, N. P. (1985b), 'A review of the Swaziland economy in 1984, projections for 1985 and beyond', *Econews*, vol. 2, no. 8. pp. 1-12
Cohen, N. P. (1985c), '1985/1986 Government of Swaziland budget and Minister of Finances budget speech', *Econews*, vol. 11. no. 7, pp. 1-8.
Cohen, N. P. & Green, T. (1984), 'Swaziland Cyclone Disaster Situation: Report Number Five', United States Agency for International Development, Mbabane.
Commonwealth Development Corporation (1976), *Annual Report and Statement of Accounts: Year Ended 31st December 1976,* London.
Commonwealth Development Corporation (1983), *Annual Report and Statement of Accounts: Year Ended 31st December 1983*, London.
Commonwealth Development Corporation (1984a), *Annual Report and Statement of Accounts: Year Ended 31st December 1984*, London.
Commonwealth Development Corporation. (1984b), 'An Introduction to Vuvulane Irrigated Farms: A Sugar Cane Based Small-Holder Settlement Scheme', Mbabane.
Conyers, D. (1982), *An Introduction to Social Planning in the Third World,* John Wiley & Sons, New York.
Cook, P. A. W. (1930), 'The Inqwala Ceremony of the Swazis', *Bantu Studies*, vol. 4. pp. 22-34.
Coulson, A. (1979), *African Socialism in Practice,* Spokesman, Nottingham.
Coulson, A. (1981), 'Agricultural Policies in Mainland Tanzania' in Heyer, J. & Williams, G. (eds.), *Rural Development in Tropical Africa*, MacMillan Press, London.
Coulson, A. (1982), *Tanzania: A Political Economy,* Clarendon Press, Oxford.
Cruise, S. E. & Holleman, J. F. (1964), 'The Statistical Frame' in Holleman, J. F. (ed.), *Experiment in Swaziland*, Oxford University Press, London.
Crush, J. (1977), 'Swaziland: Development, Dependence and Interdependence', Paper presented to The Special Session on Southern Africa: Salt Lake City Meetings of the Association of American Geographers, Salt Lake City.

Crush, J. (1980a), 'The colonial division of space: The significance of the Swaziland land partition', *International Journal of African Historical Studies*, vol. 13, no. 1, pp. 71-86.
Crush, J. (1980b), 'Modernisation and Mystification in a Southern African State', *Proceedings of the Third Annual University College of Botswana/Southern African Universities Social Science Conference,* pp.168-180, Gaborone.
Crush, J. (1986), 'Colonial Coercion and the Swazi Tax Revolt of 1903-1907' in Daniel, J, & Stephen, M. (eds.), *Historical Perspectives on the Political Economy of Swaziland*, Social Sciences Research Unit, Kwaluseni.
Daniel, J. (1979a), 'The Culture of Dependency and Political Education in Swaziland', University of Swaziland, Kwaluseni.
Daniel, J. (1979b), 'The Culture of Dependency: Introduction and Selected Bibliography', Kwaluseni: Social Sciences Research Unit/ University of Swaziland.
Daniel, J. (1981), 'The Culture of Dependency and Political Education in Africa' in Cohen, D. L. & Daniel, J. (eds.), *Political Economy of Africa,* Longman, London.
Daniel, J. (1984), 'A Comparative Analysis of Lesotho and Swaziland's Relations with South Africa', *South African Review II*, Raven Press, Johannesburg.
Daniel, J. (1986), 'Swaziland in the Context of Southern African Destabilisation' in Daniel, J, & Stephen, M. (eds.), *Historical Perspectives on the Political Economy of Swaziland*, Social Sciences Research Unit, Kwaluseni.
Daniel, J. B. McI. (1964), 'The Swazi Rural Economy' in Holleman, J. F. (ed.), *Experiment in Swaziland*, Oxford University Press, London.
Davies, R. H., O'Meara, D. & Dlamini, S. (1985), *The Kingdom of Swaziland: A Profile*, Zed Books, London.
Davis, M. J. (1980), 'The Rural Development Areas Programme', in Langford, A. N. & Martin, P. M. (eds.), *Strategies for Increasing Agricultural Production in Swaziland*, The Royal Swaziland Society for Science and Technology, Mbabane.
De Vletter, F. (1982), 'Labour migration in Swaziland: Recent trends and implications', *South African Labour Bulletin,* vol. 6/7, pp. 114-139.
De Vletter, F. (1983a), 'Introduction' in De Vletter, F. (ed.), *The Swazi Rural Homestead,* Kwaluseni: Social Sciences Research Unit.

De Vletter, F. (1983b), 'A Socio-Economic Profile of Swazi Rural Homesteads: A Summary of the Main Findings Arising from the Swaziland Rural Homestead Survey' in De Vletter, F. (ed.), *The Swazi Rural Homestead,* Social Sciences Research Unit, Kwaluseni.

De Vletter, F. (1986), *Small Holder Irrigation Projects on Swazi Nation Land,* Paper presented to the Second Seminar presented by the Land Tenure Project, Ministry of Agriculture and Cooperatives/ University of Wisconsin, Mbabane.

Dlamini, M. H. (1978), 'Forward', *Third National Development Plan 1978/79-1982/83,* Mbabane.

Dlamini, C. M. (1985), 'Changing Land Use Patterns in Lozitha-Kwaluseni/Matsapha-Logoba Area', University of Swaziland, Kwaluseni.

Dlamini, S. C. (1984), 'Income Tax Law and Swazi Custom', Paper presented to the 1984 Seminar on Income Tax Administration. Department of Taxation, Mbabane.

Dlukula. G. (1983), 'The vanishing power of the Swazi monarchy', *Pace,* Nov. pp. 18-21.

Duggan, W. & Civille, J. R. (1976), *Tanzania and Nyerere: A Study of Ujamaa and Nationhood,* Orbis Books, New York.

Engelbrecht, J. A. (1930), 'Swazi Customs Relating to Marriage, University of Stellenbosch, Stellenbosch.

Epstein, T. S. (1962), *Economic Development and Social Change in South India,* Manchester University Press, Manchester.

Epstein, T. S. (1973), *South India: Yesterday, Today and Tomorrow,* The MacMillan Press Ltd, London.

Fallers, L. A (1961), 'Are African cultivators to be called 'peasants?', *Current Anthropology,* vol. 2, pp. 108-110.

Fallers, L. A. (1965), *Bantu Bureaucracy: A Century of Political Evolution among the Basoga of Uganda,* University of Chicago Press, Chicago.

Fortes, M. & Evans-Pritchard, E. E. (1966a), *African Political Systems,* Oxford University Press, London.

Fortes, M. & Evans-Pritchard, E. E. (1966b), 'Introduction' in Fortes, M. & Evans-Pritchard, E. E. (eds.), *African Political Systems,* Oxford University Press, London.

Foster, G. M. (1965), 'Peasant society and the image of limited good' *American Anthropologist.* vol. 67, no. 2, pp. 293-315.

Foster, G. M. (1973), *Traditional Societies and Technological Change*, Harper and Row, New York.

Foster-Carter, A. (1978), 'The modes of production controversy' *New Left Review*, Jan/Feb, pp. 47-77.

Funnell, D. C. (1982a), 'Throwing money at the problem: A case study of rural development in Swaziland', *Research Papers in Geography,* no. 10, University of Sussex.

Funnell, D. C. (1982b), 'Changes in farm income and the rural development programme in Swaziland', *Journal of Developing Areas*, vol. 16, no. 2. pp. 271-290.

Funnell, D. C. (1984), 'Small Scale Irrigation on Swazi Nation Land', Paper presented to the University of Sussex, U.K.

Gluckman, M. (1950), 'Kinship and Marriage Among the Lozi of Northern Rhodesia and the Zulu of Natal' in Radcliffe-Brown, A. R. & Forde, D. (eds.), *African Systems of Kinship and Marriage,* Oxford University Press, London.

Goudie, A. S. & Price-Williams, D. (1983), *The Atlas of Swaziland,* The Swaziland National Trust Commission, Mbabane.

Gould, J. & Kolb, W. A. (1964), *A Dictionary of Social Sciences*, Collier MacMillan Canada Ltd, Toronto.

Government of Swaziland (1961), *Swaziland*, Map Series 58/SW/2, 19,000ft 6', Photo no. 022, Directorate of Overseas Surveys, London.

Government of Swaziland (1976a), *Swaziland*, Map Series. D.O.S. 435, Sheet 2631 CA (PWD no 17), Edition 6-D.O.S., Mbabane.

Government of Swaziland (1976b), 'Rural Development Areas: Project Submission', Ministry of Agriculture and Cooperatives, Mbabane.

Government of Swaziland (1978a), 'Tinkhundla: National Centres', Ministry of Local Administration, Mbabane.

Government of Swaziland, (1978b), *Third National Development Plan 1978/79 - 1982/83*, Swaziland Printing and Publishing Ltd, Mbabane.

Government of Swaziland, (1979a), *Swaziland*, Map Series, D.O.S. 1216, Sheet 2631 CA (PWD No 17), Edition 2-D.O.S., Swaziland, Mbabane.

Government of Swaziland (1979b), '1977/78 Farm Management Survey Report Number 4, Ministry of Agriculture and Cooperatives, Mbabane.

Government of Swaziland, (1982a), *Annual Statistics Bulletins*, Central Statistical Office, Mbabane.

Government of Swaziland (1982b), 'The Border Adjustment Talks between the Kingdom of Swaziland and the Republic of South Africa', Swaziland, Mbabane.

Government of Swaziland (1983a), 'Press Reports on Border Adjustment between Swaziland and the Republic', Office of the Prime Minister, Mbabane.

Government of Swaziland (1983b), *Annual Survey of Swazi Nation Land. 1982-1983*, Central Statistics Office, Mbabane.

Government of Swaziland (1984), *Annual Review of Livestock Data and Market Trends*, Ministry of Agriculture and Cooperatives, Mbabane.

Government of Swaziland (1985), *Fourth National Development Plan 1983/84 - 1987/88*, Swaziland Printing and Publishing Co. Ltd, Mbabane.

Government of Swaziland (1985b), *Farmers Census*, Ministry of Agriculture and Cooperatives, Mbabane.

Green, E. C. (1986), 'Traditional Leadership, Community Participation and Development Education' in Daniel, J, & Stephen, M. (eds.), *Historical Perspectives on the Political Economy of Swaziland*, Social Sciences Research Unit, Kwaluseni.

Guma, X. P. (1983), 'Cash Income and Expenditure' in De Vletter, F. (ed.), *The Swazi Rural Homestead*, Social Sciences Research Unit, Kwaluseni.

Guma, X. P. & Neocosmos, M. (1986), 'Some Aspects of Poverty Among Swazi Rural Homesteads', Social Sciences Research Unit. Kwaluseni.

Guma, X. P. & Simelane, V. R. (1982), 'Small Farmer Credit and Small Farmer Attitudes Towards Cooperatives', Social Sciences Research Unit, Kwaluseni.

Han, S. (1976), *The Morning Deluge*, Panther Books, St. Albans.

Harris, M. (1980), *Culture, People, Nature*, Harper & Row, New York.

Harriss, J. (1986a), *Rural Development: Theories of Peasant Economy and Agrarian Change,* Hutchinson University Library for Africa, London.

Harriss, J. (1986b), 'General Introduction' in Harriss, J. (ed.), *Rural Development: Theories of Peasant Economy and Agrarian Change*, Hutchinson University Library for Africa, London.

Herskovits, M. S. & Bascom, W, R. (1962), 'The Problem of Stability and Change in African Culture' in Bascom, W. R. & Herskovits, W. R. (eds.), *Continuity and Change in African Cultures*, Phoenix Press, London.

Higgins, B. & Higgins, J. D. (1982), 'The Reluctant Planner: An Overview of Planning in Developing Countries' in Cook, W. D. & Kuhn, T. E. (eds.), *Studies in the Management Sciences*, vol. 17, North Holland Publishing Company, New York.

Higgott, R. A. (1983), *Political Development Theory*, Croom Helm, Canberra.

Higgott, R. & Robison, R. (1986a), *South East Asia*, Routledge & Kegan Paul, Melbourne.

Higgott, R. & Robison, R. (1986b), 'Introduction' in Higgott, R. & Robison, R. (eds.), *South East Asia*, Routledge & Kegan Paul, Melbourne.

Higgott, R., Robison, R., Hewison, K. J., & Rodan, G. (1986), 'Theories of Development and Under-Development: Implications for the Study of South East Asia' in Higgott, R. & Robison, R. (eds.), *South East Asia*, Routledge & Kegan Paul, Melbourne.

Hill, P (1986), *Development Economics on Trial: The Anthropological Case for a Prosecution*, London University Press, Cambridge.

Holleman, J. F. (1962), 'Cash, Cattle or Women: A Conflict of Concepts in a Dual Economy', Paper presented to the Economic Society of the Institute of Social Research, Durban.

Holleman, J. F. (1964a), *Experiment in Swaziland*, Oxford University Press, London.

Holleman, J. F. (1964b), 'Demography: Marriage' in Holleman, J. F. (ed.), *Experiment in Swaziland*, Oxford University Press, London.

Holleman, J. F. (1964c), 'Urbanisation' in Holleman, J. F. (ed.), *Experiment in Swaziland,* Oxford University Press, London.

Holleman, J. F. (1964d), 'Addendum: Some Aspects of Agronomic Policy in Swaziland' in Holleman, J. F. (ed.), *Experiment in Swaziland,* Oxford University Press, London.

Hughes, A. J. B. (1964a), 'Demography: Religious Affiliation' in Holleman, J. F. (ed.), *Experiment in Swaziland,* Oxford University Press, London.

Hughes, A. J. B. (1964b), 'Demography: Size and Composition of Rural Homestead Groups' in Holleman, J. F. (ed.), *Experiment in Swaziland,* Oxford University Press, London.

Hughes, A. J. B. (1964c), 'Incomes and Rural Homestead Groups' in Holleman, J. F. (ed.), *Experiment in Swaziland,* Oxford University Press, London.

Humphreys, A. S. B. & Mc Clelland, C. (1964), 'Wage Employment and Labour Resources' in Holleman, J. F. (ed.), *Experiment in Swaziland,* Oxford University Press, London.

Hunter, D. E. & Whitten, P. (1977), *The Study of Cultural Anthropology,* Harper & Row, New York.

Huppert, E. L. (1983), 'Rural Health' in De Vletter, F. (ed.), *The Swazi Rural Homestead,* Social Sciences Research Unit, Kwaluseni.

Hyden, G. (1980), *Beyond Ujamaa in Tanzania: Underdevelopment and an Uncaptured Peasantry,* Heinemann, London.

Ingle, C. R. (1975), *Village to State in Tanzania: Politics of Rural Development,* Cornell University Press, London.

Jones, H. M. (1981), 'The pre-colonial distribution of the Swazi population', *African Historical Demography,* vol. 2, pp. 331-367.

Jones, B. & Dixon, M. V. (1981), *The MacMillan Dictionary of Biography,* Mary Martin, Adelaide.

Jooste, C. J. (1964), 'Demography: Characteristics' in Holleman, J. F. (ed.), *Experiment in Swaziland,* Oxford University Press, London.

Kleist, T. J. (1982), 'The Smallholder Cotton Sector in Swaziland, Social Sciences Research Unit, Kwaluseni.

Khoza, A. R. V. (1979), 'Research for Rural Development, Paper presented to the Swaziland Institute for Educational Research, Ministry of Agriculture and Cooperatives, London.

Kuper, H. (1941), 'The development of a primitive nation' *Bantu Studies,* vol. 15, pp. 339-368.

Kuper, H. (1944), 'A ritual of kingship among the Swazi' *Africa,* vol. 14, no 5, pp. 230-257.

Kuper, H. (1945), 'The marriage of a Swazi princess', *Africa,* vol. 15, pp. 145-155.

Kuper, H. (1946), 'The Swazi reaction to missions', *African Studies,* vol. 5, no 3. pp. 177-188.

Kuper, H. (1947), *An African Aristocracy,* Oxford University Press, London.

Kuper, H. (1950), 'Kinship Among the Swazi' in Radcliffe-Brown, A. R & Forde, D. (eds.), *African Systems of Kinship and Marriage,* Oxford University Press, London.

Kuper, H. (1963), *The Swazi: A South African Kingdom,* Holt, Rinehart & Winston, New York.

Kuper, H. (1964), 'The colonial situation in Southern Africa', *The Journal of Modern African Studies*, vol. 2, no 2, pp. 149-164.

Kuper, H. (1968), 'Incwala in Swaziland: of growth and kingship', *African Arts*, vol. 1, no 3, pp. 56-59.

Kuper, H. (1971), 'Colour, Categories and Colonialism: The Swazi Case' in Turner, V. (ed.), *Colonialism in Africa 1870-1960, vol. 3 Profiles of Change: African Society and Colonial Rule,* Cambridge University Press, Cambridge.

Kuper, H. (1972a), 'The language of sites in the politics of space', *American Anthropologist*, vol. 74, no 3, pp. 411-425.

Kuper, H. (1972b), 'A royal ritual in a changing political context', *Cahiers d'Etudes Africanes*, vol. 12, no 4, pp. 593-615.

Kuper, H. (1973), 'Costume and cosmology: The annual symbolism of the Ncwala', *Man*, vol. 8, no 4, pp. 613-630.

Kuper, H. (1975), 'The Supernatural' in Byaruhanga-Akiiki, A. B. T. (ed.), *Religion in Swaziland vol. 1*, University of Swaziland, Kwaluseni.

Kuper, H. (1978), *Sobhuza II: Ngwenyama and King of Swaziland,* Duckworth, London.

Lehmann, D. (1979), *Development Studies: Four Critical Studies,* Frank Cass, London.

Levin, R. (1981), *State, Multinational and Uneven Development in the Swazi Formation: The Case of Simunye,* unpublished honours dissertation in Development Studies, University of Witwatersrand.

Levin, R. (1984), 'Sugar Production and Uneven Development', Paper presented at the University of Liverpool. U. K.

Levin, R. (1985), *Hegemony and Crisis: Royal Power in Transition,* unpublished doctoral dissertation, University of Liverpool.

Levin, R. (1987a), 'Traditional Land Tenure in Swaziland: Technical Efficiency Problems of Democratic Organisation and the Value of Legalistic Classifications' in Neocosmos, M. (ed.), *Social Relations in Rural Swaziland,* Social Sciences Research Unit, Kwaluseni.

Levin, R. (1987b), 'Contract Farming in Swaziland: Peasant Differentiation and the Constraints of Land Tenure' in Neocosmos, M. (ed.), *Social Relations in Rural Swaziland,* Kwaluseni. Social Sciences Research Unit.

Leys, C. (1976), 'The 'over-developed' post colonial state; a re-evaluation', *Review of African Political Economy,* Merlin Press, London.

Low, A. (1982), 'Farmhouse theory and rural development in Swaziland', *Development Study no 23*, University of Reading, Reading.

Low, A. R. C. (1983), 'Rural Development and Farm-Homestead Economic Behaviour' in De Vletter, F. (ed.), *The Swazi Rural Homestead,* Social Sciences Research Unit, Kwaluseni.

Low, A. R. C. (1986), *Agricultural Development in Southern Africa: Farm Household Economics and the Food Crisis,* James Currey, London.

Low, A. R. C., Kemp, R. L. & Doran, M. H. (1980), 'Cattle wealth and cash needs in Swaziland: Price response and rural development implications', *Journal of Agricultural Economics*, vol. 31, no 2, pp. 614-617.

MacMillan, H. (1986) 'Swaziland: Decolonisation and the Triumph of "Tradition"' in Daniel, J. & Stephen, M. (eds.), *Historical Perspectives on the Political Economy of Swaziland,* Social Sciences Research Unit, Kwaluseni.

Mair, L. (1969), *Witchcraft*, World University Library, London.

Mair, L. (1984), *Anthropology and Development*, MacMillan Press, London.

Magagula, G. T. (1978), 'Rural Development Areas Programme', *Journal of Southern African Affairs*, vol. 3, no 4, pp. 443-470.

Magagula, G. T. (1980), 'Rural development through settlement programmes: The organisational structural and environmental problems of the Mphethseni Settlement Scheme in Swaziland', *Journal of Southern African Affairs*, vol. 6, pp. 1-25.

Magagula, G. T. (1981), 'Rural Development and Rural Urban Migration in Swaziland', Paper presented to the Fourth General Conference of the Faculties of Agriculture in Africa, Accra.

Magagula, G. T. (1982a), 'Land Tenure and Agricultural Production in Swaziland', Paper presented to The Workshop on Land Policy and Agricultural Development, National Institute of Development and Cultural Research, Gaberone.

Magagula, G. T. (1982b), 'Towards an understanding of the Concept of Rural Development in Swaziland', Social Sciences Research Unit, Kwaluseni.

Mamdani, M. (1972), *The Myth of Population Control,* Monthly Review Press, London.

Manley, M. (1986), 'Mwalimu's Democratic Legacy' *South*, February, pp. 26.

Mann, M. (1983), *The MacMillan Student Encyclopedia of Sociology,* MacMillan Press, London.

Maro, P. S. & Mlay, W. F. I. (1979), 'Decentralisation and the organisation of space in Tanzania', *Africa*, vol. 49, no 3. pp. 291-301.

Marwick, B. A. (1966), *The Swazi: An Ethnographic Account of the Natives of the Swaziland Protectorate*, Frank Cass & Co, Ltd, London.

Marwick, M. G. (1965), *Sorcery in its Social Setting: A Study of the Northern Rhodesian Cewa,* Manchester University Press, Manchester.

Mashauri, R. K. (1971), 'Leadership Structure and Functions in an Ujamaa Village: A Case Study of Gallu' in Proctor, J. H. (ed.), *Building Ujamaa Villages in Tanzania*, Tanzania Publishing House, Dar es Salaam.

Masson, J. R. (1982), 'Prehistoric Swaziland: An Archaeological Outline', Swaziland National Trust Commission, Mbabane.

Matsebula, J. S. M. (1980), *A History of Swaziland,* 2nd Edition, Longmans, Penquin, Southern Africa Ltd, Capetown.

Matsebula, J. S. M. (1983), *A Tribute to the Late His Majesty King Sobhuza II,* Websters Pty., Ltd, Mbabane.

Matsebula, J.S.M. (1983), *The King's Eye*, Maskew Miller Longmans (Pty) Ltd, Capetown.

Matthews, K. (1987), 'Squatters on Private Tenure Farms in Swaziland' in Neocosmos, M. (ed.), *Social Relations in Rural Swaziland,* Social Sciences Research Unit, Kwaluseni.

McFadden, P. (1986), 'Women and Revolution: Women Workers in Southern Africa' in Daniel, J, & Stephen, M. (eds.), *Historical Perspectives on the Political Economy of Swaziland*, Social Sciences Research Unit, Kwaluseni.

McLellan, D. (1975), *Marx*, Fontana/Collins, London.

Mercy, C. (1983), 'The Swazi Development and Savings Bank and Credit to Small Farmers in Swaziland: Three Measures of Success' in De Vletter, F. (ed.), *The Swazi Rural Homestead,* Social Sciences Research Unit, Kwaluseni.

Middleton, J. (1960), 'The Lugbara' in Richards, A. (ed.), *East African Chiefs: A Study of Political Development in Some Uganda and Tanganyika Tribes*, Faber and Faber, London.

Miller, A. (1933), *Mamisa the Warrior*, Shuter & Shooter, Pietermaritzburg.

Morss, E. R. & Gow, D. D. (1985), *Implementing Rural Development Projects: Lessons from AID and World Bank Experiences*, Westview Press, Boulder.

Morss, E. R. & Honadle, G. H. (1985), 'Differing Agendas' in Morss, E. R. & Gow, D. D. (eds.), *Implementing Rural Development Projects*, Westview Press, Boulder.

Murdoch, G. (1968), *Soil Map of Swaziland: Southern Sheet 1:125,000*, Ministry of Agriculture and Cooperatives, Mbabane.

Musoke, I. K. S. (1971), 'Building Socialism in Bukoba: The Establishment of Rugazi (Nyerere) Ujamaa Village' in Proctor, J. H. (ed.), *Building Ujamaa Villages in Tanzania*, Tanzania Publishing House, Dar es Salaam.

Mwapachu, J. V. (1979), 'Operation Planned Villages in Rural Tanzania: A Revolutionary Strategy for Development' in Coulson, A. (ed.), *African Socialism in Practice*, Spokesman, Nottingham.

Nadel, S. F. (1954), *Nupe Religion*, Routledge & Kegan Paul, London.

Nafziger, E. W. (1979), 'A Critique of Development Economics in the U.S.' in Lehmann, D. (ed.), *Development Studies: Four Critical Studies*, Frank Cass, London.

Nkambule, S. P. (1984), 'Tax in Relationship to Small Scale business in the Kingdom of Swaziland', Paper presented to the Income Tax Administration, Swaziland Institution of Management & Public Administration, Mbabane.

Neocosmos, M. (1987a), *Social Relations in Rural Swaziland*, Social Sciences Research Unit, Kwaluseni.

Neocosmos, M. (1987b), 'Introduction' in Neocosmos, M. (ed.), *Social Relations in Rural Swaziland*, Social Sciences Research Unit, Kwaluseni.

Neocosmos, M. (1987c), 'Homogeneity and Differences on Swazi Nation Land' in Neocosmos, M. (ed.), *Social Relations in Rural Swaziland*, Social Sciences Research Unit, Kwaluseni.

Neocosmos, M. (1987d), 'The Agrarian Question in Swaziland' in Neocosmos, M. (ed.), *Social Relations in Rural Swaziland*, Social Sciences Research Unit, Kwaluseni.

Neocosmos, M. & Testerink, J. (1985), 'Contract Farming in Swaziland: A Research Proposal', Paper presented to IDRC Workshop on Contract Farming and Small-Holder Outgrower Schemes. Nairobi.

Ngubane, H. (1983), 'The Swazi Homestead' in De Vletter, F. (ed.), *The Swazi Rural Homestead*, Social Sciences Research Unit, Kwaluseni.

Ntirukigwa, E. N. (1971), 'The Land Tenure System and the Building of Ujamaa Villages in Geita: A Case Study of Kalebezo' in Proctor, J. H. (ed.), *Building Ujamaa Villages in Tanzania*, Tanzania Publishing House, Dar es Salaam.

Ntshingila, M. (1984), 'Rural Women in Wage Employment', Social Sciences Research Unit, Kwaluseni.

Nyerere, J. K. (1968), *Ujamaa: Essays on Socialism,* Oxford University Press, Dar es Salaam.

Nyerere, J. K. (1969), *Nyerere on Socialism,* Oxford University Press, Dar es Salaam.

Nyerere, J. K. (1971), 'The Mwongozo TANU Party Guidelines' in Coulson, A. (ed.), *African Socialism in Practice*, Spokesman, Nottingham.

Nyerere, J. K. (1973), *Freedom and Development*, Oxford University Press, Dar es Salaam.

Nyerere, J. K. (1975), 'TANU: Ten Years After Independence' *Tanzania Notes and Records*, vol. 76, pp. 1-49.

Nyerere, J. K. (1978), *Crusade for Liberation,* Oxford University Press, Dar es Salaam.

Nyerere, J. K. (1979a), 'The Rational Choice' in Coulson, A. (ed.), *African Socialism in Practice*, Spokesman, Nottingham.

Nyerere, J. K. (1979b), 'Freedom and Development' in Coulson, A. (ed.), *African Socialism in Practice*, Spokesman, Nottingham.

Nyerere, J. K. (1979c), 'The Arusha Declaration Ten Years After' in Coulson, A. (ed.), *African Socialism in Practice*, Spokesman, Nottingham.

O'Brien. D. C. (1979), 'Modernisation, Order and the Erosion of the Democratic Ideal' in Lehmann, D. (ed.), *Development Studies: Four Critical Studies*, Frank Cass, London.

Onim, R. (1985a), 'Nyerere's pledge' *New African*, June, p. 55.

Onim, R. (1985b), 'After Nyerere, Mwinyi' *New African*, October 1985, p. 34.

Packard, R. (1986), 'Maize, Cattle and Mosquitoes: The Political Economy of Malaria Epidemics in Colonial Swaziland' in Daniel, J, & Stephen, M. (eds.), *Historical Perspectives on the Political Economy of Swaziland*, Social Sciences Research Unit, Kwaluseni.

Pakenham, T. (1979), *The Boer War*, Johnathon Ball Paperbacks, Johannesburg.

Parkin, D. J. (1972), *Palms, Wine and Witnesses,* Chandler Publishing Company, San Fransisco.

Popkin, S. L. (1979), *The Rational Peasant*, University of California Press, Berkeley.

Portch, N. (1980 'Irrigation Schemes on Swazi Nation Land' in Langford, A. N. & Martin, P.M. (eds.), *Strategies for Increasing Agricultural Production in Swaziland*, The Royal Swaziland Society for Science and Technology, Mbabane.

Poulson, K. (1983), 'Community Development in Swaziland' in De Vletter, F. (ed.), *The Swazi Rural Homestead,* Social Sciences Research Unit, Kwaluseni.

Randell, V. & Theobald, R. (1985), *Political Change and Underdevelopment: A Critical Introduction to Third World Politics,* MacMillan, London.

Robertson, A. G. (1984), *People and the State*, Cambridge University Press, Sydney.

Rudolf, L. I. & Rudolf, S. H. (1967), *The Modernity of Tradition: Political Development in India*, University of Chicago Press, Chicago.

Russell, M. (1983), 'The Rural Swazi Homestead in its Context: A Review of the Literature and Thoughts for Future Directions' in De Vletter, F. (ed.), *The Swazi Rural Homestead,* Social Sciences Research Unit, Kwaluseni.

Russell, M. (1983a), 'Boundaries and Structures in the Swazi Homestead, Social Sciences Research Unit, Kwaluseni.

Russell, M. (1983b), 'The Aged in Swaziland: Prospects for an Old Age Problem' in McDaniels, M. C. (ed.), *A Report on the Proceedings of the Seminar on the Aging*, University of Swaziland, Kwaluseni.

Russell, M. (1984), 'Beyond Remittances: The Redistribution of Cash in Swazi Society', Social Sciences Research Unit, Kwaluseni.

Russell, M. (1985a), 'Does Swaziland Have a Proletariat?' Paper presented to the University of London Postgraduate Seminar on the Societies of Southern Africa in the Nineteenth and Twentieth Centuries, University of London.

Russell, M. (1985b), 'A Report on Landlessness in Rural Swaziland', Paper presented to FAO. University of East Anglia.

Russell, M. (1986), 'High status, low pay: Anomalies in the position of women in employment in Swaziland', *Journal of Southern African Studies*, vol. 12, no 2, pp. 293-307.

Russell, M. & Ntshingila, M. (1984), 'Rural Homestead Income in Central Swaziland', Social Sciences Research Unit, Kwaluseni.

Russell, M., Mbatha, N. & Sithole, V. (1982), 'Sample Survey of Maize Growing in Swaziland', Social Sciences Research Unit, Kwaluseni.

Rycroft, D. K. (1981), *Concise SiSwati Dictionary*, J. L. van Schaik (Pty) Ltd, Pretoria.

Sahlins, M. D. (1958), *Social Stratification in Polynesia*: *American Ethnological Society Monograph, 29*, University of Washington Press, Seattle.

Sahlins, M. D. (1961), 'The segmentary lineage: An organisation of predatory expansion', *American Anthropologist*, vol. 63, pp. 322-345.

Sahlins, M. D. (1969), 'Economic Anthropology and Anthropological Economics', *Social Sciences Information,* vol. 8, no 5, pp. 13-33.

Sahlins, M. (1972), *Stoneage Economics*, Aldine, Atherton Inc, New York.

Sandee, H. (1985), 'Growth Constraints on Small Scale Selling of Vegetables and Fruit in Swaziland: Report on Fieldwork in the Greater Manzini Area', Social Sciences Research Unit, Kwaluseni.

Sansom, B. (1976), 'A Signal Transaction and its Currency' in Kapferer, B. (ed.), *Transaction and Meaning*, Institute for the Study of Human Issues, Philadelphia.

Sansom, B. (1981), 'Cash Down for Brides' in Krige, E. J & Comaroff, J. L. (eds.), *Essays on African Marriage in Southern Africa*, Juta & Co, Johannesburg.

Schapera, J. (1966), 'The Political Organisation of the Ngwato of the Bechuanaland Protectorate' in Fortes, M. & Evans-Pritchard, E. E. (eds.), *African Political Systems*, Oxford University Press, London.

Schwager, D. (1984), *Swaziland*, Mbabane: Websters (Pty) Ltd.

Scott, J. C. (1976), *The Moral Economy of the Peasant,* Yale University Press, New Haven.

Shuenyane, M. (1985), 'Nyerere steps out of the fast lane', *Pace*, October, pp. 23-26.

Sheddick, G. J. (1953), *The Southern Sotho*, Hassell, Watson & Viney, London.

Shongwe, M. V. (1983), 'A Study of Local Government in Swaziland', University of Swaziland, Kwaluseni.

Sibisi, H. (1979), 'Sociological Observation on some Aspects of Rural Development in Swaziland: Homestead Organisation', Ministry of Agriculture and Cooperatives, London.

Sibisi, H. (1980) 'Traditional Securities and the Response to Modern Economic Opportunities', Ministry of Agriculture and Cooperatives, London.

Sibisi, H. (1982), 'The Rural Base of Swazi Society', Ministry of Agriculture and Cooperatives, Mbabane.

Simelane, J. M. B. (1984), 'The Income Tax Law and the Swazi Law and Custom', Paper presented to the Income Tax Administration. Swaziland Institution of Management and Public Administration, Mbabane.

Simelane, N. (1986), 'The State and Working Class in Swaziland' in Daniel, J. & Stephen, M. (eds.), *Historical Perspectives on the Political Economy of Swaziland,* Social Sciences Research Unit, Kwaluseni.

Sithole, V. M. & Testerink, J. (1988), 'Irrigation and Food Scarcity in Swaziland: Current Status and Research Priorities', Paper presented to the Fourth Annual Conference on food Scarcity in Southern Africa, University of Zimbabwe, Harare.

Smith, M. G. (1960), *Government in Zazzau: 1800-1956,* Oxford University Press, London.

South, (1985), 'Africa: Economic Indicators for Africa' September, 1985.

Standing, H. (1987), 'Gender Relations and Social Transformation in Swaziland: Some Comments on Social Research Possibilities' in Neocosmos, M. (ed.), *Social Relations in Rural Swaziland,* Social Sciences Research Unit, Kwaluseni.

Stephen, M. (1986a), 'The Great Diamond Rush?: Or What Archaeology Tells of Swaziland Before the Swazi' in Daniel, J, & Stephen, M. (eds.), *Historical Perspectives on the Political Economy of Swaziland*, Social Sciences Research Unit, Kwaluseni.

Stephen, M. (1986b), 'Small is Beautiful: An Overview of the Swazi Economy' in Daniel, J, & Stephen, M. (eds.), *Historical Perspectives on the Political Economy of Swaziland*, Social Sciences Research Unit, Kwaluseni.

Sundkler, B. (1976), *Zulu Zion and Some Swazi Zionists,* Oxford University Press, London.

Tawney, R. H. (1938), *Religion and the Rise of Capitalism*, Penquin Books, London.

ten Raa, E. (1983a), 'A personal report on Tanzania: A troubled country with also some considerable strengths', *African Studies Association of Australia and the Pacific Newsletters,* vol. 5, no 1, pp. 7-10.

ten Raa, E. (1983b), 'Malnutrition and social organisation: A central Tanzanian perspective', *Paideuma*, vol. 29, pp. 121-137.

Testerink, J. (1984), 'Agricultural Commercialisation in Swaziland: Farmers Compared', Social Sciences Research Unit, Kwaluseni.

Testerink, J. & Russell, M. (1983), 'Mkwalini-Mahwalala Mbabane: Socio-economic Characteristics and Housing Preferences'. Report on a Sample Survey, Social Sciences Research Unit, Kwaluseni.

Testerink, J., Funnell, D. & Freund, R. (1985), 'Report on-and Review of-the 1982/83 Sample Survey of Swazi Nation Land', Social Sciences Research Unit, Kwaluseni.

Testerink-Maas, E. (1985), 'Demographic Response on Commercialisation in Agriculture: A Case Study of Swaziland', Social Sciences Research Unit, Kwaluseni.

The Employment Bureau of Africa (1986), *Swaziland Recruitment Figures 1935 - 1982*, (Piet Retief and Mankayane Offices). A Compilation of TEBA recruitment record sheets for the period 1935 - 1982.

Theodorson, G. A. & Theodorson, A. G. (1979), *A Modern Dictionary of Sociology,* Barnes & Noble Books, New York.

Tuckett, J. R. (1975), 'Vuvulane Irrigated Farms: A Report on the First Ten Years', Commonwealth Development Corporation, Mananga.

United Nations General Assembly (1984), 'Detailed Description of Needs, Project Outlines and Background Information on the Refugee Situation', Paper presented at 2nd International Conference on Assistance to Refugees in Africa, Geneva.

United States Agency for International Development (1978a), 'An Action Memorandum for the Assistant Administrator for Africa', Mbabane.

United States Agency for International Development (1978b), *R.D.A. Infrastructure Support: Project Paper (645-0068)*, Mbabane.

United States Agency for International Development (1979), *Annual Budget Submission: Swaziland*, Washington.

United States Agency for International Development (1985), *Overview of the USAID/Swaziland Programme,* Mbabane.

Vilane, J. (1986), 'Ideology and Ideological Struggle in Swaziland' in Daniel, J, & Stephen, M. (eds.), *Historical Perspectives on the Political Economy of Swaziland*, Social Sciences Research Unit, Kwaluseni.

von Freyhold, M. (1979), 'Kitumbi-Chanika and Kitumbi-Tibili: Two Ujamaa Villages that Refused to Become One' in Coulson, A. (ed.), *African Socialism in Practice*, Spokesman, Nottingham.

Vost, D. A. (1979), 'Perinatal mortality and its prevention in Swaziland', *Curationis*, vol, 2, no 3, pp. 53-54.

Webster, A. (1984), *Introduction to the Sociology of Development,* MacMillan Press, London.

Youe, C. P. (1986), 'Imperial Land Policy in Swaziland and the African Response' in Daniel, J, & Stephen, M. (eds.), *Historical Perspectives on the Political Economy of Swaziland,* Social Sciences Research Unit, Kwaluseni.

Zeigler, P. (1985), *Mountbattern: The Official Biography*, Guild Publishing, London.

Index

Achieving membership 51
Africa Cooperative Trust 27
African Development Bank 24
Age Regiment system 2
Agri-business 9, 10
Agricultural industries 2
Agricultural intensification 64
Agricultural practice 41
Agricultural production 23
Agricultural self reliance 146
Allen 5
Ancestors 114
Aristocracy 170
Armstrong 17, 66
Armstrong and Nhlapo 129, 156
Armstrong and Russell 17
Authority 52
Bafuyi 12
Bakula 206
Balimi 12
Balimi Mbamba 12
Banford 17
Bantfu Mbamba 10
Barker 23
Barnes 201
Batsakatsi 51
Beemer 14
Bhekimpi 26
Black-Michaud 10
Blocked development 18
Boer Republics 2
Boeson, Madson and Moody 205
Bohannan 109
Bonginhosi Tractor Company 27

Bonner 1, 20
Booth 1, 20, 38, 39, 155, 171, 208
Bowen 205
British 2
British administration 2
British colonial development policies 22
British colonial government 2
Bureaucracy 162

Canadian Government 24
Capetta 17
Capitalist development 202
Capitalist enterprise 189
Capitalist modes of production 17
Cartwright 15
Cattle 5, 6
Cattle breeding 53
Cattle lending arrangements 108, 109, 210
Cattle ownership 108
Centralized marketing 101
Chayanov 5
Chemical fertiliser 41
Chemical fertilisers 77
Chemical sprays 41
Chief 46, 60, 65, 71, 98, 104, 114, 121, 141, 170, 205
Chief's Court 141, 156, 157
Chiefly authority 141, 145, 195
Chiefly power 195
Chiefly spiritual power 151
Chinese 65
Chinese Agricultural Mission 27
Christian values 57
Christian work ethics 57, 60
Christianity 59, 127
Civil courts 156, 157
Class conflict 18
Class formation 18
Class solidarity 19
Class structures 19
Client to patron relationships 93
Cohen 16, 26
Cohen and Daniel 18
Cohen and Green 16
Colonial bureaucracy 214
Colonialism 140
Colonialism and blocked development 17
Commercial exchange 109
Commercial farming 23
Commercial transactions 99, 100, 107

Commoners 98, 170
Commonwealth Development
Corporation 16, 23, 28
Communal labour 95
Community work 42, 46, 89
Concessions 22
Concessions Period 2
Concrete block housing 83
Conjugal families 139, 198
Conjugal family 112, 192
Conjugal relationships 135
Conjugality 63, 75, 122
Consanguinity 114, 117
Consensus politics 50, 53, 118, 190
Contemporary Swazi society 15
Contraception 131
Contract labour 93
Cook 14
Cotton 25
Coulson 205
Coulson 1979, 1981, 1982 205
Cruise and Holleman 17, 38
Crush 20, 22
Cultivation 5
Cyclone Domoina 55, 83

Dagga cultivation 64
Daniel 17, 18
Daniel and Stephen 18
Davies 171
Davies, O'Meara and Dlamini 18
Davis 23, 24, 39
De Vletter 17, 20, 47, 95
Decision making 48
Democracy 53
Democracy and uncertainty 56
Development policy 207
Division of labour 45
Divorce 129
Dlamini 1
Dlamini clan 4, 114, 170
Dual legal systems 156
Dualism 15, 201
Duggan and Civille 205
Dzeliwe 4

EEC 24
Ekhaya 26
Electoral College 144, 159
Electricity in irrigation 55
Emacustomer (see customer) 99
Emalima 76
Engelbrecht 14

Entrepreneurs 189
Epstein 198
European Development Bank 24
European Development Fund 24
European Economic Community 28
Exchange practices 89
Ezulweni Valley 9

Fallers 201
Farm chemicals 46
Female autonomy 128
Female labour 56, 66
Fences 80
Fictive kinship 113, 117, 120, 121
Filiation 75
Finance 89
Formal transactions 100
Foster 5, 198
Foster-Carter 204
Fourth National Development Plan 24
Freehold land 22
Fund raising programmes 50
Funnell 17, 22, 23, 24, 39

Gender relations 45, 50, 56, 88, 89, 94
Gluckman 5
Goats 5
Gogo 5
Gould and Kolb 207
Grass strips 79
Grazing areas 69
Great Councillor 170
Green 17
Guma 17
Guma and Simelane 17

Handicrafts 13
Headman 4, 5
Headmen 12, 170
Headmen committed 67
Headmen uncommitted 67
Herskovits and Bascom 201
Higgott 15, 17, 202, 204, 206, 212
Higgott 211
Higgott and Robison 15, 17, 208
Hill 198
Holleman 16, 17, 107
Homestead 4
Homestead economy 6
House of Representatives 159

House property complex 5, 63, 75, 91, 122
House/property complex 4
Hughes 17, 57
Humphreys and McClelland 17
Huppert 17
Hybrid maize 78
Hybrid seed 77
Hyden 205
Hynd 173

Imisumphe 56
Incwala 2
Independence 2
Indhlu 4
Indirect rule 2
Individual Tenure Land 6, 23, 24, 69, 174, 204, 214
Ingle 205
Inner Council 155, 170
Intensive Rural Development Area 38
Intermediate homestead 125
Intermediate homesteads 136
Inyoni Yami Scheme 28
Irrigation 41
Irrigation schemes 13, 47

Jones 2
Jooste 17
Judicial authority 117, 140, 165

Khoi-san 2
Khonta 4
Khoza 173
Kin based labour 91, 93, 109
Kin relationships 91
King Mswati III 49
Kinship 117, 121, 190
Kinship groups 114
Kliest 17
Kraal manure 41, 77
Kuper 1, 5, 14, 23, 49, 50, 57, 63, 76, 98, 102, 105, 113, 114, 115, 122, 136, 149, 152, 156, 170, 172, 173, 212
Kwaluseni 9

Labour 76, 89, 90, 92, 192
Labour migration 13
Land 140, 147, 165
Land Management Committee 104
Land Management Committees 64, 68, 71, 90, 148, 150

Land partition 38
Land Partition Proclamation 38
Land scarcity 63, 65, 71, 191
Land tenure 63
Large Assembly 144
Leadership 49
Levin 18, 28
Levirate 6
Light industry 9
Lineage rights 112
Lineages 114
Liqoqo 4
Loan indiscipline 108
Lobamba 2
Lobola 6
Lobola (see marriage cattle) 91
Local Contingent 144
Local government personnel 170, 184, 197
Locality 112, 114
LOME agreements 24
Lome Conventions 28
Lord Milner 2
Low 17, 19
Low, Kemp and Doran 19, 20
Low veldt 71
Lozitha 9
MacMillan 19
Magagula 17
Mair 64
Maize production 25
Makhositive 4
Malaria 18
Malkerns 9
Malkerns Agricultural Research Station 46
Mamdani 5
Manley,, 1986 205
Mann 207
Manzini 9
Marginal homestead 71
Marginal homesteads 102
Marginal land 66, 67, 69, 71, 90
Market transactions 99, 193
Market woman 99, 101
Market women 99, 130
Marketing 104
Marketing strategies 45
Markets 13, 185, 203
Maro and Mlay 206
Marriage cattle 6, 105, 106, 107
Marriage relations 112
Marwick 5, 14, 57, 98, 115, 122, 127
Marwick 64

Mashauri 206
Masson 2
Material culture 37
Matsapha 9
Matsebula 1, 2, 15, 172, 173, 174
Matthews 19
Mbabane 2, 9
McFadden 18
Mercy 17
Mfanasibili 172
Mfecane 2
Mhlume Sugar Company 28
Middlemen 99, 101, 104, 168
Migration planned 178
Migration unplanned 178
Miller 64
Mineral processing 2
Modern homestead 125
Modern homesteads 136
Modernity 191, 202, 214
Modes of production 18
Monogamous marriage 122, 192
Morss and Gow 212
Morss and Honadle 212
Motane 13
Motane irrigation scheme 53
Mswati III 4, 52, 168, 170, 173
Musoke 206
Muti murder see ritual murder 64
Mwapachu 1979, p.116 205
Mzimneni Resettlement Scheme 53

Nadel 201
Nafziger 15
National Association of Swazi
Women 132
National Council of Negro Women
26
National Development Corporation
of Swaziland 27
Native Area 38
Native Areas 22
Native Lands Scheme 22
Native Lands Scheme (of 1954) 39
Ndabazabantu 209
Ndwandwe 2
Neocosmos 18
Neocosmos and Testerink 19
New African 15
Ngubane 20, 204
Nguni 1, 2
Ngwane National Liberatory
Congress 173
Ngwempisi 12, 13, 22
Ngwempisi Rural Development Area

Ngwempisi Rural Development Area 21, 24
Ngwenyama 2
Nhlabatsi 175
Nkhotfotjeni 172
Nkwalini-Mahwalala 9
Norms of agnation 189
Ntimane 13, 14
Ntimane Irrigation Scheme 35, 40
Ntirukigwa 206
Ntshingila 17
Nyerere 205

O'Brien 15
Onim,, 1985a, 1985b 205

Packard 18
Pakenham 2
Parkin 189
Pastoralism 5
Patrilineal descent 72
Patrilocality 72
Patronage 173, 174
Peasant economy 5
Pedi 107
Plot ownership 52
Political control 115
Political paternalism 52, 191
Political rights 56
Polygynous marriage 122
Polygyny 58, 70, 83, 198, 204
Population growth 65
Poulson 16
Power 112
Power and authority 165
Preferential membership 52
Primogeniture 75, 118
Proprietorial control 75, 97, 122, 190, 191, 200, 215
Proprietorial rights 63, 75, 76, 83, 90, 109, 191

Quality control 101

Randall and Theobald 207, 208
Randell and Theobald 15, 17
Real Farmer 84, 128
Real Farmers 6, 12, 13, 29, 48, 63, 65, 71, 73, 76, 89, 94, 98, 112, 117, 170, 189, 198, 211, 214
Real Farmers and democracy 52

Real People 10, 12, 13, 21, 49, 169, 170, 183, 205
Real Person 73
Reciprocity 93
Regional Assembly 141, 158
Regional Council 163
Republic of China 24
Resident Commissioner 2
Right of usufruct 215
Rights in genetricem 105
Rights in uxorem 105
Rights of filiation 105
Rights of usufruct 190, 192
Ritual exchanges 99
Ritual murder 64, 155
Ritual transactions 105
Robertson 15, 208
Royal hegemony 19
Royal patronage 171
Royal Swazi Airlines 27
Ruling aristocracy 170, 171, 172, 195, 201, 203, 207, 211, 212, 214
Rural commoners 170, 183, 197
Rural Development Area Extension Officer 46
Rural Development Areas Programme 1, 15, 21, 22, 23, 26, 27, 38, 52, 53, 60, 63, 65, 81, 115, 121, 141, 146, 148, 162, 165, 166, 185, 201, 205, 206, 211, 214
Rural Development Areas-Infrastructure Support (RDA-IS) Project 24
Rural homestead 14
Rural self reliance 22
Rural wage earners 178, 180, 197
Russell 17, 19, 204
Russell and Ntshingila 17
Russell, Mbatha and Sithole 20
Rycroft 10

Sahlins 5, 198
Salary earners 176, 196
Sandee 100, 101
Sangoma 103
Sansom 107
Schapera 5
Schwager 15
Scott 5, 199
Second Boer War 2, 22
Semi-commercial exchange 109

Semi-commercial farming 23
Semi-commercial transactions 99, 100
Shangaan 93
Sheddick 5
Sheep 5
Sheunyane 205
Sheunyane,, 1985, pp.23-26). 205
Shona 107
Shongwe 158, 159, 160
Sibisi 10, 20
Sidlanis 179
Simelane 18, 19
Simunye Sugar 28
Sisa 6
SiSwati Development Board 174
Sithole and Testerink 47, 48
Sive 4
Small Assembly 144
Small holder schemes 23
Smith 201
Sobhuza I 2
Sobhuza II 2, 49, 52, 55, 140, 146, 155, 160, 172, 174, 191, 207, 211, 212
Social differentiation 91, 169, 195, 216
Social mobility 174
Social obligations 101
Social organization 1, 4, 52
Social sanctions 51
Social Sciences Research Unit 17
Sororal polygyny 6
Sotho 2
South Africa 13, 18, 38, 207
South Africa and KaNgwane 63
South African goldmines 6
South African mines 175
Southern African Development Co-ordination Conference 24
Sozisa 4
Standing 18
State corporatism 208
Status 112
Status of women 132, 139, 193
Stephen 18
Structural functionalism 14
Structural-functionalism 19
Subsistence production 9
Sugarcane 23
Sundkler 57
Swazi Defense Force 171
Swazi Electricity Board 54
Swazi elite 18
Swazi Industrial Development Corporation 27

Swazi Life 15
Swazi nation 207
Swazi Nation Land 6, 12, 22, 23, 24, 25, 26, 67, 108, 145, 171, 174, 178, 183, 201, 210, 214
Swazi National Assembly 160, 162, 170
Swazi National Church 57
Swazi National School 173
Swazi Observer 26, 48
Swazi Savings and Development Bank 27
Swazi state 200, 207
Swazi tradition 15
Swaziland 1
Swaziland Development and Savings Bank 108
Symbolic power 141
Systems of production 1, 4

Tawney 57
Taxation 203
Telecommunications 9
Ten Raa 205
Testerink 17
Testerink and Russell 9, 17
Testerink, Funnell and Freund 17
Testerink-Maas 17
The status of women 129
Tibiyo Taka Ngwane 28, 171, 203, 214
Times of Swaziland 24, 26, 63, 64
Tinkhundla 46
Tisuka Taka Ngwane 171
Tobacco 25
Tourism 15
Tractor hire 50
Tradition 16, 191, 202, 214
Traditional 140
Traditional authority 56, 189
Traditional healers 151
Traditional homestead 98, 122, 147
Traditional labour roles 98
Traditional maize seed 77
Traditional maize varieties 78
Traditional power 140
Traditional rights of usufruct 191
Traditional social organization 9
Traditional subsistence agriculture 23
Traditional Swazi society 6
Transport system 9
Transvaal 2
Tuckett 16, 23

U.S. Embassy 46
Ujamaa 205
Umkhaya 86
Umuti 4
Uneven development 18
Unitarian Service Committee of Canada 46
Unitarian Services Committee of Canada 26
United Nations 24
University of South Africa 173
University of Swaziland 9, 17
Urban elite 10, 172, 173, 195, 196, 201, 214
Urban elites 170
Urban patron 177
Urban wage earners 178, 179, 196
Urban wage work 67
USAID 16, 24, 38, 48, 53
Usufruct 6, 66, 75, 90, 115
Usufruct rights of 63

Vegetable growing 41
Vegetable marketing 42
Vegetables 25
Vilane 18, 19
Von Freyhold 205
Von Wissell 173
Vost 16
Vuvulane 16
Vuvulane Irrigated Farms 23

Wage and salary earners 10, 170, 175, 196, 201
Wage earners 176
Wage labour 93
Wage work 180
Watchtower Sect 59
Webster 15
West German International Development Agency 27
Witch hunt 153
Witchcraft 51, 103, 115, 141, 152, 153, 154, 190, 192
Witchcraft allegations 154
Witchcraft in politics 53
Witchhunt 151
Women 5, 13, 18, 45, 56, 75, 90, 94, 107, 128, 178
Women and power 95
Women and reproduction 129
Women and work 129
Women managers 87
Women's labour 66

Women's work 193
Work ethics 48, 58
Work party 136, 137, 138
Workpractices 45
World Bank 24, 26

Youe 19

Zionism 58, 59, 154
Zionism and monogamy 85
Zionist Churches of South Africa
57